WILDLIFE IN AIRPORT ENVIRONMENTS

Wildlife Management and Conservation

Paul R. Krausman, *series editor*

WILDLIFE
in Airport Environments

Preventing Animal–Aircraft Collisions
through Science-Based Management

EDITED BY

Travis L. DeVault,
Bradley F. Blackwell
& Jerrold L. Belant

Published in Association with *THE WILDLIFE SOCIETY*

THE JOHNS HOPKINS UNIVERSITY PRESS | BALTIMORE

© 2013 The Johns Hopkins University Press
All rights reserved. Published 2013
Printed in the United States of America on acid-free paper
9 8 7 6 5 4 3 2 1

The Johns Hopkins University Press
2715 North Charles Street
Baltimore, Maryland 21218-4363
www.press.jhu.edu

Library of Congress Cataloging-in-Publication Data

Wildlife in airport environments : preventing animal-
aircraft collisions through science-based management /
edited by Travis L. DeVault, Bradley F. Blackwell, and
Jerrold L. Belant.
        pages cm. — (Wildlife management and
conservation)
    Includes bibliographical references and index.
    ISBN-13: 978-1-4214-1082-1 (hardcover : alk. paper)
    ISBN-13: 978-1-4214-1083-8 (electronic)
    ISBN-10: 1-4214-1082-6 (hardcover : alk. paper)
    ISBN-10: 1-4214-1083-4 (electronic)
    1. Airports—Bird control. 2. Wildlife pests—Control.
I. DeVault, Travis L., 1974– editor of compilation.
II. Blackwell, Bradley F., 1962– editor of compilation.
III. Belant, Jerrold L., 1963– editor of compilation.
    TL725.3.B5W55 2013
    363.12'47—dc23   2013004817

A catalog record for this book is available from the British
Library.

*Special discounts are available for bulk purchases of this book.
For more information, please contact Special Sales at 410-516-
6936 or specialsales@press.jhu.edu.*

The Johns Hopkins University Press uses environmentally
friendly book materials, including recycled text paper that
is composed of at least 30 percent post-consumer waste,
whenever possible.

Contents

Contributors

Michael L. Avery, U.S. Department of Agriculture, Wildlife Services, National Wildlife Research Center

Jerrold L. Belant, Mississippi State University

Kristin M. Biondi, Mississippi State University

Bradley F. Blackwell, U.S. Department of Agriculture, Wildlife Services, National Wildlife Research Center

Jonathon D. Cepek, U.S. Department of Agriculture, Wildlife Services

Larry Clark, U.S. Department of Agriculture, Wildlife Services, National Wildlife Research Center

Tara J. Conkling, Mississippi State University

Scott R. Craven, University of Wisconsin–Madison

Paul D. Curtis, Cornell University

Travis L. DeVault, U.S. Department of Agriculture, Wildlife Services, National Wildlife Research Center

Richard A. Dolbeer, U.S. Department of Agriculture, Wildlife Services

David Felstul, U.S. Department of the Interior

Esteban Fernández-Juricic, Purdue University

Alan B. Franklin, U.S. Department of Agriculture, Wildlife Services, National Wildlife Research Center

Sidney A. Gauthreaux Jr., Clemson University

Michael Lavelle, U.S. Department of Agriculture, Wildlife Services, National Wildlife Research Center

James A. Martin, Mississippi State University

Rebecca Mihalco, U.S. Department of Agriculture, Wildlife Services

Paige M. Schmidt, U.S. Fish and Wildlife Service

Thomas W. Seamans, U.S. Department of Agriculture, Wildlife Services, National Wildlife Research Center

Kurt C. VerCauteren, U.S. Department of Agriculture, Wildlife Services, National Wildlife Research Center

Brian E. Washburn, U.S. Department of Agriculture, Wildlife Services, National Wildlife Research Center

Preface

On 15 January 2009, the world learned—in dramatic fashion—that wildlife pose serious hazards to aircraft. On that day, US Airways Flight 1549, an Airbus 320 carrying 155 people, made an emergency landing in the Hudson River in New York City after ingesting Canada geese (*Branta canadensis*) into both engines at an altitude of ~2,900 feet (880 m) following takeoff from LaGuardia Airport (Marra et al. 2009, National Transportation Safety Board 2010). Historically, most people had never considered the extent of hazards posed to aircraft by birds and other wildlife. After all, how can birds, which generally weigh a few kilograms at most, bring down an airliner? Don't they just bounce off or get shredded by the powerful engines?

The Flight 1549 incident brought about widespread awareness of wildlife–aircraft collisions (also called wildlife strikes or bird strikes), which has been welcomed by biologists, airport managers, and other personnel who manage wildlife at airports and worked to develop solutions to this problem for several decades. Reporting of wildlife strikes increased following Flight 1549 (reporting is voluntary in the USA; Dolbeer 2009), and in many cases more resources have been allocated to management activities and research efforts. Increased awareness of the wildlife-strike problem also has ushered in a new wave of devices, materials, and services designed to reduce the risk of wildlife strikes; these range from grass-seed mixtures intended to deter foraging by Canada geese to sophisticated avian radars. Some of these products are quite promising, whereas others might not mitigate risk any better than vehicle-mounted "deer whistles" (Valitzski et al. 2009). In our efforts to reduce strike risk, how can we discern the effective tools from the ineffective ones? Where should we direct our research and development efforts in the future?

As with any technical challenge, we must rely on science. Effective management of wildlife in airport environments, like all types of wildlife damage management, is based on principles from wildlife ecology, physiology, and behavior (Conover 2002). By considering how these disciplines interact in the airport context, we can better understand how and why animals respond to various mitigation methods (at both the individual and population levels), learn why and under what conditions some management tools and techniques work better than others, and more intelligently direct our future research and management efforts. To that end, this book provides a broad review of tools and techniques used to prevent wildlife collisions with aircraft, focusing on the science underlying the methods. Readers interested in a "how-to" guide for airport wildlife management should further consult MacKinnon (2004) and Cleary and Dolbeer (2005).

We begin this book with an introductory chapter summarizing the history of wildlife strikes with aircraft and organize the remainder into three parts. In the first, Wildlife Management Techniques, six chapters cover wildlife deterrents (visual, chemical, tactile, and auditory), exclusion methods, translocation strategies, and population management. The second part, Managing Resources (four chapters), begins with two chapters that discuss food and water resources, two key wildlife attractants that are present at nearly all airports. Part II continues with a chapter on managing turfgrass, a dominant land cover (and wildlife attractant) at airports across the world. Chapter 11 considers alternatives to turfgrass at airports and the proper role of airports in wildlife conservation. In Part III, Wildlife Monitoring, we present an overview of animal movements followed by two more specialized chapters on avian radar and avian survey methods at airports. We conclude the book with a discussion on directions for future research.

The alert reader will notice a few themes that emerge from these chapters. First, there is an emphasis on managing for the most hazardous wildlife (i.e., those species most likely to cause aircraft damage when struck; DeVault et al. 2011) present at a particular airport. Although some management activities are considered common sense and good practice at all airports (e.g., covering trash containers and discouraging the deliberate feeding of wildlife), others are more specific to context, such as the various types of vegetation management. Care should always be taken that the elimination of one species does not inadvertently create an attractant for another, more hazardous species. No single strategy or tool will be developed that successfully mitigates wildlife strikes in every location for every species; therefore an approach that combines techniques in an integrated fashion is most effective (Conover 2002). The wide array of topics covered in this book underscores the importance of using an integrated approach to managing wildlife at airports. Finally, although we focus on wildlife management at the airport (where most strikes occur), we also recognize the importance of reducing strikes at higher altitudes, outside the airport environment. Strikes at higher altitudes are more likely to cause substantial damage than strikes at lower altitudes (Dolbeer 2006) and are increasing in frequency (Dolbeer 2011). Some of the techniques we discuss, such as aircraft lighting that elicits earlier alert and escape behaviors by birds in response to oncoming aircraft, are promising for reducing strikes at higher altitudes, away from the airport environment.

Without question, modern airports face many demands. They must promote safety above all else, but airports are increasingly considered to be drivers of local economies, promoters of "green" energy production and other environmental initiatives, and, at times, sites

for the conservation of rare species (Blackwell et al. 2009, DeVault et al. 2012). The demands confronting airports are expected to intensify as air traffic increases and as airport infrastructures change to meet increased capacity. These necessary changes must consider how wildlife—particularly those species posing the greatest strike hazards—is managed. It is our hope that this book will help airport managers, biologists, airport and urban planners, students, consultants, businesspeople, and others understand how effective wildlife management at airports contributes to the safety and efficiency of air travel worldwide.

Literature Cited

Blackwell, B. F., T. L. DeVault, E. Fernández-Juricic, and R. A. Dolbeer. 2009. Wildlife collisions with aircraft: a missing component of land-use planning for airports. Landscape and Urban Planning 93:1–9.

Cleary, E. C., and R. A. Dolbeer. 2005. Wildlife hazard management at airports: a manual for airport personnel. Second edition. Federal Aviation Administration, Office of Airport Safety and Standards, Washington, D.C., USA.

Conover, M. R. 2002. Resolving human–wildlife conflicts: the science of wildlife damage management. CRC Press, Boca Raton, Florida, USA.

DeVault, T. L., J. L. Belant, B. F. Blackwell, J. A. Martin, J. A. Schmidt, L. W. Burger Jr., and J. W. Patterson Jr. 2012. Airports offer unrealized potential for alternative energy production. Environmental Management 49:517–522.

DeVault, T. L., J. L. Belant, B. F. Blackwell, and T. W. Seamans. 2011. Interspecific variation in wildlife hazards to aircraft: implications for airport wildlife management. Wildlife Society Bulletin 35:394–402.

Dolbeer, R. A. 2006. Height distribution of birds recorded by collisions with civil aircraft. Journal of Wildlife Management 70:1345–1350.

Dolbeer, R. A. 2009. Trends in wildlife strike reporting, part 1: voluntary system, 1990–2008. Report DOT/FAA/AR-09/65. U.S. Department of Transportation, Federal Aviation Administration, Washington, D.C., USA.

Dolbeer, R. A. 2011. Increasing trend of damaging bird strikes with aircraft outside the airport boundary: implications for mitigation measures. Human–Wildlife Interactions 5:235–248.

MacKinnon, B. 2004. Sharing the skies: an aviation industry guide to the management of wildlife hazards. TP 13549. Transport Canada, Aviation Publishing Division, Ottawa, Ontario, Canada.

Marra, P. P., C. J. Dove, R. A. Dolbeer, N. F. Dahlan, M. Heacker, J. F. Whatton, N. E. Diggs, C. France, and

G. A. Henkes. 2009. Migratory Canada geese cause crash of US Airways Flight 1549. Frontiers in Ecology and the Environment 7:297–301.

National Transportation Safety Board. 2010. Loss of thrust in both engines after encountering a flock of birds and subsequent ditching on the Hudson River, US Airways Flight 1549, Airbus A320-214, N106US. Aircraft Accident Report NTSB/AAR-10/03. Washington, D.C., USA.

Valitzski, S. A., G. J. D'Angelo, G. R. Gallagher, D. A. Osborn, K. V. Miller, and R. J. Warren. 2009. Deer responses to sounds from a vehicle-mounted sound-production system. Journal of Wildlife Management 73:1072–1076.

Acknowledgments

The U.S. Department of Agriculture's National Wildlife Research Center (NWRC); U.S. Department of Transportation, Federal Aviation Administration (FAA); U.S. Department of Defense (DoD); and numerous representatives from private industry provided funding for much of the research reported herein. However, opinions expressed in this book do not necessarily reflect current FAA or DoD policy decisions governing the control of wildlife at or near airports. The FAA and NWRC also provided funding support for the preparation of this book, and NWRC and Mississippi State University provided salary support for the editors. We thank Elizabeth Poggiali (NWRC) for spending many hours proofing chapters, Vincent Burke (The Johns Hopkins University Press) for his editorial assistance, Ashleigh McKown for copyediting the manuscript, The Wildlife Society for its support of this book, and the many airport biologists around the world for their work that keeps us safe.

We are also pleased to recognize the time and comments offered by the chapter reviewers listed below. Our intent in obtaining reviews was to select at least two referees for each chapter, including a research biologist or professor and a wildlife manager. We thank all the reviewers for sharing their expertise.

Chapter Reviewers

James Armstrong
Auburn University

Ken Ballinger
Arkion Life Sciences

Tara Baranowski
U.S. Department of Agriculture,
    Wildlife Services

Thomas Barnes
University of Kentucky

Scott Barras
U.S. Department of Agriculture,
    Wildlife Services

Scott Beckerman
U.S. Department of Agriculture,
    Wildlife Services

Jason Boulanger
Cornell University

Frank Boyd
U.S. Department of Agriculture,
    Wildlife Services

Peter Coates
U.S. Geological Survey, Western
    Ecological Research Center

Richard Dolbeer
U.S. Department of Agriculture,
    Wildlife Services (retired)

Joelle Gehring
Michigan State University

Allen Gosser
U.S. Department of Agriculture,
    Wildlife Services

Robert Kennamer
University of Georgia

Tommy King
U.S. Department of Agriculture,
    Wildlife Services, National
    Wildlife Research Center

Matt Klope
U.S. Navy

Martin Lowney
U.S. Department of Agriculture,
    Wildlife Services

James Martin
Mississippi State University

Mark McConnell
Mississippi State University

Terri Pope
Utah Division of Wildlife Resources

Craig Pullins
U.S. Department of Agriculture, Wildlife Services

Russ Reidinger
Lincoln University

Gene Rhodes
University of Georgia

Sam Riffell
Mississippi State University

Laurence Schafer
U.S. Department of Agriculture, Wildlife Services

Peter Scott
Indiana State University

Thomas Seamans
U.S. Department of Agriculture, Wildlife Services,
    National Wildlife Research Center

Jason Suckow
U.S. Department of Agriculture, Wildlife Services

Mark Tobin
U.S. Department of Agriculture, Wildlife Services,
    National Wildlife Research Center

Guiming Wang
Mississippi State University

Brian Washburn
U.S. Department of Agriculture, Wildlife Services,
    National Wildlife Research Center

Scott Werner
U.S. Department of Agriculture, Wildlife Services,
    National Wildlife Research Center

Gary Witmer
U.S. Department of Agriculture, Wildlife Services,
    National Wildlife Research Center

WILDLIFE IN AIRPORT ENVIRONMENTS

1

RICHARD A. DOLBEER

The History of Wildlife Strikes and Management at Airports

The first human-powered flight took place in December 1903, when Orville and Wilbur Wright successfully flew their experimental aircraft at Kitty Hawk, North Carolina, USA. Birds, which had been practicing powered flight for about 150 million years, suddenly had a new "competitor" for airspace, and the bird–aircraft collision problem (hereafter referred to as bird strikes) began shortly thereafter (Cleary and Dolbeer 2005). On 7 September 1905, the first reported bird strike, as recorded by Orville Wright in his diary, occurred when his aircraft hit a bird over a cornfield near Dayton, Ohio, USA. Flocks of red-winged blackbirds (*Agelaius phoeniceus*) and other birds are often attracted to cornfields in autumn to feed (Dolbeer 1990), making it likely that a red-winged blackbird caused the first known bird strike. In addition to birds, mammals and other wildlife can be a problem for safe aircraft operations. The first reported mammal strike occurred on 25 July 1909, at the start of Louis Bleriot's historic first flight across the English Channel from Les Baraques, France. While warming up the engine of the Bleriot XI aircraft, an excited farm dog ran into the spinning propeller (http://www.pbs.org/wgbh/nova/transcripts/3207_bleriot.html).

On 3 April 1912, Calbraith Rodgers, the first person to fly across the continental USA, was killed in the first fatal crash resulting from a wildlife strike when his aircraft struck a gull (Laridae) along the coast of Southern California (Cleary and Dolbeer 2005). Despite this tragic event, strikes with birds and other wildlife were of little concern for the first 50 years of aviation.

In fact, only three civil aircraft were destroyed and two human fatalities were documented worldwide between 1912 and 1959 (Fig. 1.1). But in October 1960, a turboprop-powered Lockheed Electra crashed in Boston Harbor, Boston, Massachusetts, USA, shortly after takeoff, following the ingestion of over 200 European starlings (*Sturnus vulgaris*) into the air intakes of three of the aircraft's four engines. Sixty-two people died, a fatality count which to date remains the highest for a bird-induced plane crash. During 1960–2010, bird and other wildlife strikes destroyed 160 civil aircraft, 49 from 2001 through 2010. For military aviation, more destroyed aircraft and deaths related to wildlife strikes occurred in the 1940s due to the introduction of jet-powered aircraft and increased numbers of low-level flights.

Why So Many Wildlife Strikes?

There are multiple reasons for the dramatic increase in wildlife strikes since the 1960s. First, the advent of turbine-powered passenger aircraft in the 1960s revolutionized air travel, but it also magnified the problem of wildlife strikes. Early piston-powered commercial aircraft were noisy and relatively slow. Birds could usually avoid these aircraft, and those strikes that did occur typically resulted in little or no damage to the plane. However, modern jet aircraft are faster than their predecessors, relatively quiet, and their engine fan blades are often more vulnerable to strike damage than propellers. When turbine-powered aircraft collide with birds or other wildlife, structural damage affecting the

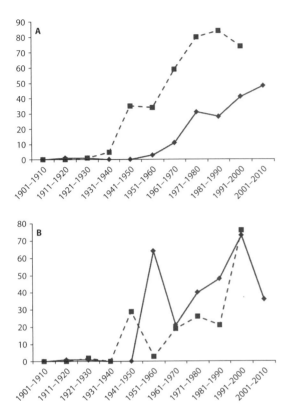

Fig. 1.1. (A) Number of aircraft destroyed and (B) human fatalities by bird and other wildlife strikes by decade. Solid lines show data for civil aircraft, and dashed lines show data for military aircraft. The years 2001–2010 are not included for military aircraft because the data for that decade are incomplete. Data from Richardson and West (2000), Thorpe (2003, 2005, 2010), and Dolbeer et al. (2012)

integrity and function of the engine or flight surface is more likely (Dolbeer et al. 2012).

Second, multiple-engine damage from the ingestion of flocks of birds became a growing concern as commercial air carriers replaced older three- or four-engine aircraft fleets with more efficient and quieter two-engine turbine-powered aircraft (Frings 1984, Hovey et al. 1992). About 90% of the 2,100 U.S. passenger aircraft had three or four engines in 1965. In 2005, the passenger fleet in the USA had grown to about 8,200 aircraft, and only about 10% had three or four engines (U.S. Department of Transportation 2009). With steady advances in technology over the past several decades, today's two-engine aircraft are more powerful and reliable than yesterday's three- and four-engine aircraft. However, in the event of a multiple-ingestion event (as

exemplified by the US Airways Flight 1549 incident on 15 January 2009; National Transportation Safety Board 2010), aircraft with two engines have vulnerabilities not shared by their three or four engine–equipped counterparts (Solman 1973). In addition, birds appear less able to detect and avoid modern jet aircraft with quieter turbofan engines compared to older aircraft with noisier engines (Solman 1976; Burger 1983; Kelly et al. 1999, 2001; Kelly and Allan 2006; see also International Civil Aviation Organization 1993). Modern turbofan engines typically have inlets with larger diameters than earlier jet-powered aircraft, which also increases the probability of bird ingestion (Banilower and Goodall 1995).

Third, worldwide air travel has become commonplace. Data from the Federal Aviation Administration (FAA) indicate that commercial air traffic in the USA increased from about 14 million movements (takeoffs or landings) in 1975 to 25 million movements in 2010 (FAA 2010). Worldwide, commercial jet aircraft movements increased from about 26 million in 1991 to 40 million in 2010 (Boeing Commercial Airplanes 2010). Aircraft have also assumed a vital role in tactical and logistical military operations. These factors have resulted in dramatically increased air traffic (Kelly and Allan 2006).

Fourth, the increased use of the skies by traveling humans has coincided with an unprecedented period of successful wildlife management and environmental protection in North America and elsewhere in the world. Aggressive natural resource and environmental protection programs by public and private wildlife management and conservation groups beginning in the late 1960s have contributed to impressive population increases of many large-bodied species such as white-tailed deer (Odocoileus virginianus), American alligators (Alligator mississippiensis), Canada geese (Branta canadensis), double-crested cormorants (Phalacrocorax auritus), sandhill cranes (Grus canadensis), American white pelicans (Pelecanus erythrorhynchos), gulls (Larus spp.), raptors (falcons, hawks, and eagles; order Falconiformes), vultures (Cathartes aura and Coragyps atratus), and wild turkeys (Meleagris gallopavo; Buurma 1996, Dolbeer and Eschenfelder 2003). At the same time, many of these species (e.g., white-tailed deer, Canada geese, and wild turkeys) have expanded into suburban and urban areas, including airports, and are thriving in response to protection and changes to habitats in these areas (Smith et al. 1999). Most of these species have body masses >1.8 kg (4 lb) and thus are more likely than smaller species to cause damage to aircraft when struck, and exceed certification standards for most airframe components and engines (Dolbeer et al. 2000, 2012; Dolbeer and Eschenfelder 2003; DeVault et al. 2011). Thus the increased probability of damaging wildlife strikes since the 1960s is primarily related to the increase in air traffic by two-engine, large-inlet, turbine-powered aircraft concurrent with major increases in populations of many large-bodied wildlife species.

Mitigating Risk through Wildlife Management Programs

The previously mentioned Lockheed Electra crash in Boston Harbor in 1960 marked the dawn of wildlife management programs to mitigate bird strikes in airport environments. Initially, leadership in this emerging field came from Canada and Europe, as exemplified by the creation of Bird Strike Committee Canada and Bird Strike Committee Europe (now the International Bird Strike Committee, or IBSC) in the 1960s. At that time, researchers sought to collect bird-strike statistics in Europe and North America. In the early 1970s, research was published on vegetation management at British airports to discourage starlings and other bird species (Brough 1971), and a biologist with the Canadian Wildlife Service wrote the first book outlining the nature and management of the bird-strike problem (Blokpoel 1974).

The bird-induced crashes of a Learjet 24 at DeKalb-Peachtree Airport, Atlanta, Georgia, USA, in 1973 and a DC-10 at John F. Kennedy International Airport, New York, New York, USA, in 1974 (Thorpe 2005) were both attributed, at least in part, to nearby landfills that attracted blackbirds (Icteridae) and gulls. These crashes led to recommended land-use restrictions near airports by the FAA and International Civil Aviation Organization. In addition, civil aviation authorities developed regulations (e.g., FAA 2004) to require that airports experiencing bird strikes assess and manage these hazards through habitat management and control techniques. The FAA in 1991 and the International Civil Aviation Organization in 2008 expanded their regulations and standards to include hazardous terrestrial wildlife such

as deer (Dolbeer et al. 2005, International Civil Aviation Organization 2009).

In 1991, a major program to manage the local nesting gull population was launched at John F. Kennedy International Airport (Dolbeer et al. 1993), which marked the initiation of aggressive management actions at airports to mitigate risks of bird and other wildlife strikes in the USA. During the 1990s, the FAA and International Civil Aviation Organization developed major databases on such strikes (Dolbeer et al. 2012). These databases indicated that most damaging strikes caused by birds in the 1990s (about 65% of strikes with civil aircraft in the USA) were in the airport environment (<152 m [500 feet] above ground level; Dolbeer 2006), which reinforced efforts to develop effective wildlife hazard management programs at airports (e.g., Cleary and Dolbeer 2005). Transport Canada published a sequel to Blokpoel's (1974) book in 2004 (MacKinnon 2004). From 2005 through 2006, the FAA developed standards for biologists working at airports (FAA 2012) and the IBSC developed a set of best practices for bird control units at airports (Allan 2005).

As a result of these efforts by federal agencies, private-sector biologists, and airport operational personnel, there has been a steady increase in the implementation and improvement of wildlife hazard management plans for airports worldwide over the past 20 years. For example, biologists from the U.S. Department of Agriculture Wildlife Services program provided assistance at 832 airports to mitigate wildlife risks during 2010, compared to only 42 and 193 airports assisted in 1990 and 1998, respectively (Begier and Dolbeer 2011; Fig. 1.2, see p. 4). An analysis of strike data for civil aviation in the USA from 1990 through 2009 indicated that these airport-based programs reduced the rate of damaging strikes at airports (Dolbeer 2011), but likely had little or no impact on the rate of damaging strikes outside the immediate airport environment (>152 m above ground level).

The Future

Although measurable progress has been made in recent years to keep hazardous birds off airports (Dolbeer 2011), increased efforts are needed to make areas within and surrounding airports less attractive to these same birds (e.g., de Hoon and Buurma 2000, Washburn

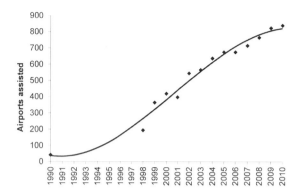

Fig. 1.2. Number of U.S. civil and military airports assisted (including through technical and direct management) by the U.S. Department of Agriculture, Animal and Plant Health Inspection Service, Wildlife Services, to reduce wildlife hazards (1990–2010). Data from Begier and Dolbeer (2011)

2010). In addition, airport managers worldwide face new challenges regarding the management of wildlife hazards. As the demand for air travel has increased, forthcoming changes to airport capacity are being met with calls for planning to maintain biodiversity in the airport environment (Blackwell et al. 2009a). Further, concerns over fossil fuel consumption have fostered research in renewable energy, with airport properties serving as potential sites for solar, biofuel, and (under limited circumstances) wind energy production (Blackwell et al. 2009a, DeVault et al. 2012). How changes in airport capacity and land use will ultimately affect wildlife populations and the associated risks to aviation (e.g., DeVault et al. 2011, Martin et al. 2011) remains unclear.

Programs to manage wildlife and associated habitats at and near airports will not, by themselves, resolve this conflict. To mitigate the risks caused by birds within and outside airport fences, increased efforts are needed in the field testing and refinement of bird-detecting radar systems (Nohara et al. 2005, Klope et al. 2009; Chapter 13). The ultimate goal is to integrate bird-detecting radar into air traffic control (ATC) procedures in a manner analogous to what has been accomplished with wind-shear detection and avoidance. These efforts will require increased risk management training for flight crews, air carrier operations personnel, and ATC personnel (Eschenfelder and DeFusco 2010). In addition, more research is needed on avian sensory perception and reaction to moving objects. Such research may lead to the development of aircraft lighting systems (which could include various pulse rates and wavelengths in the electromagnetic spectrum) to enhance detection, speed perception, and avoidance of departing and arriving aircraft by birds (Blackwell et al. 2009b, 2012).

The mitigation of risks posed to aviation by birds and other wildlife is a complex endeavor in today's world, requiring expertise from a variety of biological, engineering, and safety disciplines. The following chapters discuss various components of the conflict between nature and aviation, as well as the research and management efforts underway to make our skies safer for birds and people. Progress is being made on several fronts, but much remains to be done.

LITERATURE CITED

Allan, J. R. 2005. Minimum best practice standards for aerodrome bird control. Proceedings of the International Bird Strike Committee 27:1–8.

Banilower, H., and C. Goodall. 1995. Bird ingestion into large turbofan engines. DOT/FAA/CT-93/14. Federal Aviation Administration Technical Center, Atlantic City, New Jersey, USA.

Begier, M. J., and R. A. Dolbeer. 2011. Protecting the flying public and minimizing economic losses within the aviation industry: technical, operational, and research assistance provided by USDA-APHIS-Wildlife Services to reduce wildlife hazards to aviation, fiscal year 2010. U.S. Department of Agriculture, Animal and Plant Health Inspection Service, Wildlife Services, Washington D.C., USA.

Blackwell, B. F., T. L. DeVault, E. Fernández-Juricic, and R. A. Dolbeer. 2009a. Wildlife collisions with aircraft: a missing component of land-use planning on and near airports? Landscape and Urban Planning 93:1–9.

Blackwell, B. F., T. L. DeVault, T. W. Seamans, S. L. Lima, P. Baumhardt, and E. Fernández-Juricic. 2012. Exploiting avian vision with aircraft lighting to reduce bird strikes. Journal of Applied Ecology 49:758–766.

Blackwell, B. F., E. Fernández-Juricic, T. W. Seamans, and T. Dolans. 2009b. Avian visual configuration and behavioural response to object approach. Animal Behaviour 77:673–684.

Blokpoel, H. 1974. Bird hazards to aircraft. Canadian Wildlife Service, Ministry of Supply and Services, Ottawa, Ontario, Canada.

Boeing Commercial Airplanes. 2010. Statistical summary of commercial jet aircraft accidents, 1959–2010. http://www.boeing.com/news/techissues/pdf/statsum.pdf.

Brough, T. 1971. Experimental use of long-grass in the U.K. Proceedings of the Bird Strike Committee Europe 6.

Burger, J. 1983. Jet aircraft noise and bird strikes: why more birds are being hit. Environmental Pollution 30:143–152.

Buurma, L. S. 1996. Superabundance of birds: trends, wetlands and aviation. Proceedings of the Bird Strike Committee Europe 23:43–50.

Cleary, E. C., and R. A. Dolbeer. 2005. Wildlife hazard management at airports: a manual for airport operators. Second edition. Federal Aviation Administration, Office of Airport Safety and Standards, Washington, D.C., USA.

de Hoon, A., and L. Buurma. 2000. Influence of land use on bird mobility: a case study of Eindhoven Airport, 1998–1999. Proceedings of the International Bird Strike Committee 25.

DeVault, T. L., J. L. Belant, B. F. Blackwell, J. A. Martin, J. A. Schmidt, and L. Wes Burger Jr. 2012. Airports offer unrealized potential for alternative energy production. Environmental Management 49:517–522.

DeVault, T. L., J. L. Belant, B. F. Blackwell, and T. W. Seamans. 2011. Interspecific variation in wildlife hazards to aircraft: implications for airport wildlife management. Wildlife Society Bulletin 35:394–402.

Dolbeer, R. A. 1990. Ornithology and integrated pest management: the red-winged blackbird (Agelaius phoeniceus) and corn. Ibis 132:309–322.

Dolbeer, R. A. 2006. Height distribution of birds as recorded by collisions with civil aircraft. Journal of Wildlife Management 70:1345–1350.

Dolbeer, R. A. 2011. Increasing trend of damaging bird strikes with aircraft outside the airport boundary: implications for mitigation measures. Human–Wildlife Interactions 5:235–248.

Dolbeer, R. A., J. L. Belant, and J. Sillings. 1993. Shooting gulls reduces strikes with aircraft at John F. Kennedy International Airport. Wildlife Society Bulletin 21:442–450.

Dolbeer, R. A., and P. Eschenfelder. 2003. Amplified bird-strike risks related to population increases of large birds in North America. Proceedings of the International Bird Strike Committee 26:49–67.

Dolbeer, R. A., S. E. Wright, and E. C. Cleary. 2000. Ranking the hazard level of wildlife species to aviation. Wildlife Society Bulletin 28:372–378.

Dolbeer, R. A., S. E. Wright, and P. Eschenfelder. 2005. Animal ambush at the airport: the need to broaden ICAO standards for bird strikes to include terrestrial wildlife. Proceedings of the International Bird Strike Committee 27:102–113.

Dolbeer, R. A., S. E. Wright, J. Weller, and M. J. Begier. 2012. Wildlife strikes to civil aircraft in the United States, 1990–2011. Serial Report No. 17 DOT/FAA/AS/00-6(AAS-310). U.S. Department of Transportation, Federal Aviation Administration. Washington D.C., USA.

Eschenfelder, P., and R. DeFusco. 2010. Bird strike mitigation beyond the airport, pilots must be prepared for bird strike avoidance and damage control. AeroSaftey World. August:44–47.

FAA. Federal Aviation Administration. 2004. Title 14 U.S. Code of Federal Regulations. Part 139: certification of airports. U.S. Department of Transportation, Washington, D.C., USA.

FAA. Federal Aviation Administration. 2010. Terminal Area Forecast (TAF) system. http://aspm.faa.gov/main/taf.asp.

FAA. Federal Aviation Administration. 2012. Qualifications for wildlife biologist conducting wildlife hazard assessments and training curriculums for airport personnel involved in controlling wildlife hazards on airports. Advisory Circular 150/5200-36. U.S. Department of Transportation, Washington, D.C., USA.

Frings, G. 1984. A study of bird ingestions into large high bypass ratio turbine aircraft engines. DOT/FAA/CT-84/13. U.S. Department of Transportation, Federal Aviation Administration, Atlantic City, New Jersey, USA.

Hovey, P. W., D. A. Skinn, and J. J. Wilson. 1992. Engine bird ingestion experience of the Boeing 737 aircraft—expanded database (October 1986–September 1989). DOT/FAA/CT-91/32. U.S. Department of Transportation, Federal Aviation Administration, Atlantic City, New Jersey, USA.

International Civil Aviation Organization. 1993. Convention on international civil aviation (international standards and recommended practices). Annex 16: environmental protection. Third edition. Montreal, Quebec, Canada.

International Civil Aviation Organization. 2009. Convention on international civil aviation (international standards and recommended practices). Annex 14: aerodromes. Volume I. Aerodrome design and operations. Fifth edition. Montreal, Quebec, Canada.

Kelly, T. C., and J. Allan. 2006. Ecological effects of aviation. Pages 5–24 in J. Davenport and J. L. Davenport, editors. The ecology of transportation: managing mobility for the environment. Springer, Dordrecht, Netherlands.

Kelly, T. C., R. Bolger, and M. J. A. O'Callaghan. 1999. The behavioral response of birds to commercial aircraft. Proceedings of the Bird Strike Committee–USA/Canada 99:77–82.

Kelly, T. C., M. J. A. O'Callaghan, and R. Bolger. 2001. The avoidance behaviour shown by the rook (Corvus frugeilegus) to commercial aircraft. Pages 291–299 in H. J. Pelz, D. P. Cowan, and C. J. Feare, editors. Advances in vertebrate pest management. Filander Verlag, Fürth, Germany.

Klope, M. W., R. C. Beason, T. J. Nohara, and M. J. Begier. 2009. Role of near-miss bird strikes in assessing hazards. Human–Wildlife Interactions 3:208–215.

MacKinnon, B. 2004. Sharing the skies: an aviation industry guide to the management of wildlife hazards. TP 13549. Transport Canada, Aviation Publishing Division, Ottawa, Ontario, Canada.

Martin, J. A., J. L. Belant, T. L. DeVault, L. W. Burger Jr., B. F. Blackwell, S. K. Riffell, and G. Wang. 2011. Wildlife risk to aviation: a multi-scale issue requires a multi-scale solution. Human–Wildlife Interactions 5:198–203.

National Transportation Safety Board. 2010. Loss of thrust in both engines after encountering a flock of birds and subsequent ditching on the Hudson River, US Airways Flight

1549, Airbus A320-214, N106US. Aircraft Accident Report NTSB/AAR-10/03. Washington, D.C., USA.

Nohara, T. J., P. Weber, A. Premji, C. Krasnor, S. A. Gauthreaux, M. Brand, and G. Key. 2005. Affordable avian radar surveillance systems for natural resource management and BASH applications. Proceedings of the International Radar Conference. IEEE, 14–18 August 2005, Vancouver, British Columbia, Canada.

Richardson, W. J., and T. West. 2000. Serious birdstrike accidents to military aircraft: updated list and summary. Proceedings of the International Bird Strike Committee 25:67–98.

Smith, A., S. R. Craven, and P. D. Curtis. 1999. Managing Canada geese in urban environments. Jack Berryman Institute Publication 16. Cornell Cooperative Extension, Ithaca, New York, USA.

Solman, V. E. F. 1973. Birds and aircraft. Biological Conservation 5:79–86.

Solman, V. E. F. 1976. Aircraft and birds. Proceedings of the Bird Control Seminar 7:83–88.

Thorpe, J. 2003. Fatalities and destroyed aircraft due to bird strikes, 1912–2002. Proceedings of the International Bird Strike Committee 26:85–113.

Thorpe, J. 2005. Fatalities and destroyed aircraft due to bird strikes, 2002–2004 (with an appendix of animal strikes). Proceedings of the International Bird Strike Committee 27:17–24.

Thorpe, J. 2010. Update on fatalities and destroyed civil aircraft due to bird strikes with appendix for 2008 to 2009. Proceedings of the International Bird Strike Committee 29:1–9.

U.S. Department of Transportation. 2009. National transportation statistics. Table 1-13: active U.S. air carrier and general aviation fleet by type of aircraft research and innovative technology administration. http://www.bts.gov/publications/national_transportation_statistics/html/table_01_13.htm.

Washburn, B. E. 2010. Evaluation of the North Shore Marine Transfer Station and its compatibility with respect to bird strikes and safe air operations at LaGuardia Airport. Report for the Associate Administrator of Airports, Office of Airport Safety and Standards, Airport Safety and Operations. Federal Aviation Administration. Washington, D.C., USA.

PART I · WILDLIFE MANAGEMENT TECHNIQUES

2

Bradley F. Blackwell
Esteban
Fernández-Juricic

Behavior and Physiology in the Development and Application of Visual Deterrents at Airports

In the first major treatise on the science of wildlife damage management, Conover (2002) dedicated a short review of visual stimuli used to deter wildlife from specific areas or resources. The brevity of the review reflects the fact that these techniques have traditionally been developed over short periods and used to confront an immediate problem, generally through trial and error. Because humans perceive visual stimuli differently than other animals (Schwab 2012), deterrents based on human perception likely fall short in saliency of the stimuli (i.e., how well the stimuli stand out against a background). However, assessment of visual stimuli (both deterrents and cues) in the context of animal sensory physiology and behavior holds promise for the development of novel and more effective methods to mitigate negative human–wildlife interactions.

As Conover (2002) noted, visual deterrents are generally intended to provoke a fear response. Examples include scarecrows or other human forms, object movement (e.g., Mylar tape; Dolbeer et al. 1986), predator models (Conover 1982, 1985; Conover and Perito 1981), animal effigies (Avery et al. 2002, Seamans 2004), methods that provoke neophobia (e.g., coyote [*Canis latrans*] response to novel objects; Windberg 1997), and methods that combine movement and neophobia (e.g., use of lasers in bird dispersal [Blackwell et al. 2002, Gorenzel et al. 2002] and fladry against wolves [*C. lupus*; Musiani et al. 2003]). We can also include the use of border collies (*C. familiaris*) against birds at airports (Sodhi 2002) and other

dogs to protect livestock against mammalian predators (Rondinini and Boitani 2007) and contact with wild ungulates (Gehring et al. 2010). Visual stimuli that cue alert responses, as opposed to provoking fear, have been investigated relative to deer–vehicle collisions (D'Angelo et al. 2006, Blackwell and Seamans 2009) and in eliciting desired behavioral responses in birds (e.g., avoiding collision with static objects [Martin 2011]; enhancing detection and response to approaching aircraft [Blackwell and Bernhardt 2004; Blackwell et al. 2009, 2012a; Fernández-Juricic et al. 2011]). In this chapter we distinguish between visual methods that serve as the primary deterrent or cue and color cues used as conditioned stimuli in the context of chemical repellents (Chapter 3). We refer to visual stimuli intended to provoke fear and cues designed to enhance detection of objects as *visual deterrents*.

The immediate and long-term effectiveness of visual deterrents varies by species (e.g., Koehler et al. 1990, Mason 1998), season, group size (Dolbeer et al. 1986), habitat, and even legal constraints (Conover 2002). Moreover, the effectiveness of visual deterrents (or lack thereof) targeting birds or mammals is inherently linked to detectability, discriminability, and memorability, the three factors that govern design of animal signals (Guilford and Dawkins 1991; see also Endler 1992). In this initial section of the chapter, unless otherwise cited, we relate the discussion of these factors to Guilford and Dawkins (1991).

The environment through which the signal is transmitted, the sensory capabilities of the receiving animal, and the cognitive processing of the receiver affect detectability of natural signals. Signal discriminability connotes that the receiver recognizes in the signal stimuli some category by which a particular behavior (e.g., flight, avoidance of a prey item) is warranted. Memorability of the signal is linked to learning, where the signal is eventually associated with a particular outcome. From the perspective of natural signals, Guilford and Dawkins (1991) assigned these three factors as components of strategic signal design and tactical design, or of signal efficacy.

Whereas strategic design in animal signaling is concerned with whether or why (in terms of fitness) the receiver responds appropriately, the second component—efficacy—affects the probability that the signal will reach its target destination and elicit a response. For instance, a signal might be salient because it is both easily detectable and occurs within the context of familiar habitat (e.g., coyote response to an intruder's sign or a novel object within the animal's territory; see Windberg 1997). Taken another way, if this same coyote encountered a novel object outside its territory, though the object is readily visible, it would likely show little interest simply because there is not an apparent intent (i.e., fitness consequence) to the signal.

Efficacy in animal signaling also entails aspects of what Guilford and Dawkins (1991) termed the receiver's "psychological landscape," or the cognitive processing of the signal that takes place behind the immediate sensory organs (in this case the eyes). Specifically, an animal's signal might be composed of cues important within one context but intended for another. For example, Guilford and Dawkins (1991) note that peacock (*Pavo cristatus*) tail coloration (i.e., the eye spots) will draw the immediate attention of the peahen from a vigilance perspective, but divert her attention to indicators of male fitness.

Unquestionably, signaling and signal reception by animals are multifaceted and complex (Endler 1990, 1992; Endler and Théry 1996). For a visual deterrent to effectively communicate some a priori risk to a target animal or cue that attracts the animal's attention, we must ask what traits or conditions are required for detectability, both to reinforce signal strength and to extend the period of effectiveness. Our purpose in the

remainder of this chapter is to discuss the importance of visual physiology, behavior, and ecological context as components in the design and effective use of visual deterrents against mammals and birds. We review briefly (1) vision in mammals and birds relative to other sensory paths and deterrent efficacy, (2) threat recognition in animals, and (3) how visual deterrents are currently used in the airport environment, as well as the efficacy of these methods.

Vision as a Sensory Pathway

Vertebrates have two types of photoreceptors: rods and cones (McIlwain 2006, Schwab 2012). Rods deal with dim conditions and are not activated by bright light. The ability to perceive color is dependent on the number of different visual pigments present in the cone photoreceptors (Cuthill 2006). Animals with a single visual pigment cannot perceive color but can determine differences in brightness of a signal (Land and Nilsson 2002). Animals with two or more visual pigments can perceive color. This capability is explained by the way the visual system works. Light entering the retina stimulates the visual pigments of the photoreceptors to different degrees (depending on the wavelength distribution of light and the peak sensitivity of the visual pigments). The visual system at the retinal level (amacrine cells, bipolar cells, horizontal cells, ganglion cells) uses stimulation ratios (instead of absolute stimulation values of a given visual pigment type) to estimate how much each photoreceptor is stimulated compared to the others, and then sends this information to the visual centers in the brain (Land and Nilsson 2002). Color perception is based on these stimulation ratios, which will vary depending on the number of visual pigments in the retina and the wavelength peak sensitivity of each visual pigment (Gouras 2007). Animals with two visual pigments are known as dichromats, three visual pigments as trichromats, four visual pigments tetrachromats, and so on.

Mammals

The relevance of the visual systems of mammals varies widely across taxa (e.g., Langley 1983, and references therein) because some species, such as ungulates, rely more on olfaction and hearing than on vision. However, dogma often suggests limitations to ungulate vision that are inaccurate. For example, white-tailed deer (*Odocoileus virginianus*) and fallow deer (*Dama dama*) are not color-blind. Both species are dichromatic with peak sensitivity at 450–460 nm (i.e., "blue") and at 537 nm (i.e., "green"). Their eyes also contain rod cells (up to 90% of the photoreceptors in the retina) that are activated only under dim conditions (Jacobs et al. 1994; see also VerCauteren and Pipas 2003, Warren et al. 2008). In addition, like other mammals active at night, ungulates possess a tapetum lucidum (Dukes 1969, D'Angelo et al. 2008) that reflects incidental light back through the retina and associated photopigments a second time, further enhancing vision in dim light.

The visual capability of deer at night is not necessarily limited to changes in brightness or intensity of the stimuli, however, but is dependent upon the intensity and spectra (i.e., wavelengths that compose the hue or color) of ambient lighting (e.g., presence of street lighting) and the time of exposure (Jacobs et al. 1994, VerCauteren and Pipas 2003, D'Angelo et al. 2008). As we alluded above, detection of a signal is but one component of signal effectiveness. Attempts to exploit signal detection in white-tailed deer via roadside, wavelength-specific light cues (i.e., those visually detectable by deer) synchronized with vehicle approach at night lead ironically to an increased chance of deer–vehicle collisions because of the confusion caused by the visual cues (D'Angelo et al. 2006). In contrast, Blackwell and Seamans (2009) showed that vehicle-based lighting that is more visually detectable to white-tailed deer enhanced detection and response to approaching vehicles. We suspect the difference in responses of deer in these studies is due to the application or saliency of light signals (roadside versus on the vehicle) relative to the potential threat (the vehicle). Importantly, however, findings from both studies highlight the interplay of brightness and spectra relative to deer response to reflected and direct vehicle lighting.

As with ungulates, the importance of the visual pathway to canids also varies with species and context (e.g., dominance of vision over other senses depends on ambient lighting; Langley 1983). Jacobs et al. (1993) examined the visual pigments of the domestic dog, is-

land gray fox (*Urocyon littoralis*), red fox (*Vulpes vulpes*), and Arctic fox (*Alopex lagopus*). According to the authors, besides rods, each of the four genera are (like ungulates) dichromatic, sharing one cone cell with peak absorption at ~555 nm ("green") and a second cone cell with peak absorption from 430 to 435 nm ("blue"). We note, however, that there is little evidence that canids necessarily respond to color signals, whereas movement and novelty have played more of a role in deterrent efficacy (e.g., Windberg 1997, Mason 1998, Musiani et al. 2003) than considerations for spectral sensitivity of the target animal.

Birds

Unlike most mammals, vision represents a primary sensory pathway for birds (Walls 1942) and is highly developed, as evidenced by the relative size of the eyes to the skull (in some species the combined weight of the eyes exceeds that of the brain; Sillman 1973). Further, birds have visual systems that differ substantially from mammalian vision, including higher temporal visual resolution and sensitivity in a broader range of the spectrum (e.g., ~370–700 nm; Cuthill 2006, Martin 2011, Fernández-Juricic 2012). Birds are tetrachromats, with four types of visual pigments in their cone photoreceptors, and species differ at the level of the visual pigment sensitive to shorter wavelengths (Hart and Hunt 2007). Some species are ultraviolet sensitive (with a visual pigment peaking at 355–380 nm) or violet sensitive (with the visual pigment peaking at 402–426 nm). The peak sensitivity of the three other visual pigments in birds varies as follows: short-wavelength sensitive (427–463 nm), medium-wavelength sensitive (499–506 nm), and long-wavelength sensitive (543–571 nm). Additionally, birds have organelles within each cone photoreceptor known as oil droplets; these organelles filter light before it reaches the visual pigment, thereby enhancing color discrimination (Cuthill 2006). Birds also have rods, but in diurnal species they amount to about 20% of the photoreceptors (Querubin et al. 2009).

Sillman (1973) contended that no treatment of the biology of birds is sufficient without consideration of vision. The effective development and use of visual deterrents against birds must also consider the complexity of their visual systems, as well as the context of the stimulus relative to the desired behavioral response (Blackwell 2002).

Common Properties of Mammalian and Avian Visual Systems

Despite the aforementioned differences, the visual systems of mammals and birds share a characteristic that can have implications for the development of visual deterrents. Both taxa process visual information in similar ways (see Dowling 2012).

Photoreceptors are responsible for converting optical information into a neural signal. Several photoreceptors are generally connected to a single retinal ganglion cell (which transfers information from the retina to the brain through the optic nerve) via different cells (amacrine, bipolar, horizontal). The group of photoreceptors that connect to a given ganglion cell forms a receptive field. Receptive fields of adjacent ganglion cells overlap in such a way that a given receptive field (on center) is surrounded by another receptive field (off center). When light hits the on-center receptive field, the associated ganglion cell is stimulated. When light hits the off-center receptive field, the associated ganglion cell is stimulated. However, when light simultaneously hits both the on- and off-center receptive fields, both ganglion cells inhibit each other, decreasing the cell's firing rate. The bottom line is that mammalian and avian visual systems at the retinal level work on the basis of differences in stimulation between center and surrounding receptive fields, rather than absolute changes in light intensity. This means the visual system is tuned to how much a given object reflects light relative to the light reflected from the background (Land and Nilsson 2002), rather than the absolute properties of the object (e.g., total amount of light reflected or wavelength reflected).

One implication of how the visual system processes visual information is that we cannot establish how color or light intensity is perceived by other species, because the number and properties of light-sensitive components of the retina (e.g., visual pigments, oil droplets; Cuthill 2006) influence the perceptual experience. These light-sensitive components will influence the stimulation ratios that an object (e.g., signal) and its visual background generate on the retina and, consequently, the visual contrast response. Visual deterrents

will be processed in a similar way. However, we can theoretically estimate the visual saliency of a deterrent for a given visual system.

Mathematical models (Vorobyev and Osorio 1998, Endler and Mielke 2005, Montgomerie 2006), used with freely available software (such as AVICOL; see Gomez 2006), can estimate visual contrast (i.e., chromatic, based on hues, and achromatic, based on brightness). These visual contrast models provide an estimate of how much an object stands out from the background. The required information to parameterize the models includes the reflectance of the object, reflectance of the visual background, and irradiance (spectral properties of the ambient light), which can be measured with an off-the-shelf spectrometer. Additionally, visual contrast models require empirical information on visual parameters of the target species (or a related species), metrics that are currently available in the literature, including sensitivity of the visual pigments and oil droplets (Hart and Hunt 2007) and relative density of cone photoreceptors (Hart 2001). These models can be used to establish the color and brightness that would enhance the visual contrast of a deterrent for a given species, assuming that the most salient deterrents for a given visual system could enhance an avoidance response. This assumption can be tested empirically through behavioral experiments. Overall, this sensory approach to develop visual deterrents can narrow the range of visual deterrents that have the highest chance, due to their visual saliency, of triggering a desired behavioral response.

Innate versus Learned Responses to Signals

Following our discussion of how we can enhance the saliency of a visual deterrent by better understanding the sensory system of the target species, it is logical to ask whether the characteristics of the deterrent are inherently meaningful. In other words, do the characteristics of the deterrent have the potential to stimulate innate avoidance or antipredator behaviors (Caro 2005), as with some natural signals? Or will the stimulus require a period of learning accompanied by reinforcement via other stimuli, for example, enhancing apparent predation risk to ring-billed gulls (*Larus delawarensis*; Conover 1987) or American crows (*Corvus brachyrhynchos*; Marzluff et al. 2010)? Inglis and

Isaacson (1984) demonstrated that exposure of wood-pigeon (*Columba palumbus*) wing marks is aversive to conspecifics, and that these marks might serve as a natural visual alarm (see also Murton 1974). In contrast, Shivik et al. (2003) noted that use of disruptive visual stimuli (e.g., fladry) against wolves can decrease predation, but does not produce or stimulate an aversion to the resource.

Also, natural signal colors from potential prey, such as warning-colored or aposematic prey (Poulton 1890, Guilford 1990, Gamberale-Stille and Guilford 2003) or other food resources (Herrera 1985, Altshuler 2001, Honkavaara et al. 2004), seem to be adapted for producing maximal differences in stimulation of avian photoreceptors (Finger and Burkhardt 1994, Vorobyev et al. 1998), serving as cues that stimulate innate or learned responses to the resource. Innate avoidance by birds of aposematic patterns characteristic of potential vertebrate prey is common (e.g., Rubinoff and Kropach 1970; Smith 1975, 1977; Caldwell and Rubinoff 1983). Innate and learned avoidance of aposematic invertebrate prey (e.g., yellow and black banding patterns) by birds has also been demonstrated (Schuler 1982, Schuler and Hesse 1985, Lindström et al. 1999). Any exploitation of behavioral responses to aposematic coloration for deterrent design must also consider that the primary context for application would likely entail deterrence of foraging, as opposed to provoking a sense of fear. Findings by Avery et al. (1999) with regard to bird avoidance of certain seed colors hold promise for the development of seed coatings to deter bird predation of newly seeded crops. Similarly, color treatments might also reduce avian mortality due to consumption of pesticide-treated baits or seeds (e.g., de Almeida et al. 2010).

In the context of antipredator behavior and our ability to exploit these behaviors, particularly salient visual signals from predators include aspects of size, shape, and movement pattern (e.g., Tinbergen 1948, Blumstein et al. 2000, Veen et al. 2000, Goth 2001; see also Inglis and Isaacson 1984). These same visual signals are also important in learned antipredator responses (e.g., Marzluff et al. 2010) and in response to novel threats. Chamois (*Rupicapra r. rupicapra*) in the Swiss Alps fled the approach of paragliders (possibly perceived as raptors because of flight dynamics) by as much as 900 m (2,953 feet; Schnidrig-Petrig and Ingold 2011). Similar escape behaviors in response to

the presence of aircraft have been observed in other mammals, such as mountain sheep (*Ovis canadensis*) disturbance by helicopters (Bleich et al. 1994) and hauled-out ringed seal (*Phoca hispida*) disturbance by fixed-wing aircraft and helicopters (Born et al. 1999). Necropsies and examination of associated injuries of birds struck by aircraft indicated that antipredator responses occurred before collision (Bernhardt et al. 2010). The efficacy of visual deterrents intended to elicit a fear response is also linked to similar predator traits (e.g., Boag and Lewin 1980, Avery et al. 2002, Seamans 2004), and there is potential to enhance the risk perceived by an animal relative to unnatural stimuli (e.g., Ydenberg and Dill 1986, Frid and Dill 2002, Stankowich and Blumstein 2005).

Visual Deterrents at Airports

Cleary and Dolbeer (2005:111–135) provide the most current review of control techniques, including visual deterrents, available for use at airports. These techniques include the use of natural predators such as trained falcons (Blokpoel 1976) or dogs, both of which have gained popularity in recent years because they are intended as nonlethal management approaches. In addition, Mylar flagging for short-term applications against birds, predator and prey effigies, and handheld lasers continue to be used at airports. Here we examine in greater detail the use of handheld lasers, effigies, and more recent advances in visual deterrents.

Lasers

A common application in the use of lasers against birds stems from findings by Blackwell et al. (2002) related to marked avoidance responses by captive Canada geese (*Branta canadensis*) to a moderate-power, 650-nm laser (Fig. 2.1).

However, Blackwell et al. (2002) also noted that wavelength sensitivity does not connote deterrence. They cited research reporting long-wavelength sensitivity in European starlings (*Sturnus vulgaris*), rock pigeons (*Columba livia*), and mallards (*Anas platyrhynchos*), yet captive groups of these species exhibited no avoidance or only a limited response to treatment from moderate-power, 630-nm (starlings) and 650-nm lasers (all three species). Ambient conditions or context

Fig. 2.1. Captive Canada geese moving perpendicular to (away from) a laser beam from a moderate-power (i.e., 5–500 mW), 650-nm laser in experiments conducted by Blackwell et al. (2002). Laser power specifications from U.S. Department of Labor, Occupational Health and Safety Administration; see http://www.osha.gov/dts/osta/otm/otm_iii/otm_iii_6.html. Photo credit: Bradley F. Blackwell

(e.g., captive versus free-ranging birds, light conditions surrounding a roost that can affect dark adaption by retinal photopigment and subsequent sensitivity to laser beams and beam spots, or predation risk outside a roost) likely affect potential responses to laser treatment. Sherman and Barras (2004) found that Canada goose response to a 650-nm laser was limited by ambient lighting and pond size. Similarly, Gorenzel et al. (2002) found that American crows occupying urban roosts responded to moderate-power, 630- and 650-nm lasers, but quickly reoccupied roosts. In an evaluation of moderate-power, 473- and 534-nm lasers against white-tailed deer, VerCauteren et al. (2006) noted that deer detected laser treatments, but the devices were ineffective as dispersal tools. In reference to findings by Blackwell et al. (2002) relative to birds, the authors noted that differential effectiveness of lasers may be due to species-specific differences in threat perception and avoidance behavior.

Effigies

The effectiveness of effigies in eliciting a desired behavioral response (e.g., area avoidance or flight) is inherently linked to spectral and form attributes, presentation, movement, and context. Seamans (2004) found that a hanging (as opposed to supine) taxidermy mount of a turkey vulture (*Cathartes aura*), susceptible to movement by wind and in full view of roosting vultures, resulted in abandonment of a roost used during fall migration (Fig. 2.2).

In another study, Avery et al. (2002) used carcasses and taxidermy mounts of turkey and black vultures (*Coragyps atratus*) to disperse mixed roosts of vultures from communication towers. But in a test of a floating, mold-injected plastic Canada goose effigy as an area repellent against territorial pairs of Canada geese during late summer, Seamans and Bernhardt (2004) found no effect. Similarly, mold-injected plastic raptor models failed to deter European starlings from nest boxes (Belant et al. 1998). Unlike taxidermy mounts that were natural in appearance and form, as well as positioned such that erratic movement could occur, the plastic effigies lacked one or both of these attributes.

In contrast, Mason et al. (1993) deterred snow geese (*Chen caerulescens*) from agricultural fields via white plastic flagging, a cue typically used by hunters to decoy geese. In this case, however, systematic placement of the flagging, versus clumped placement used during hunting seasons, likely contributed to the deterrent effect. Effigies in the form of duck decoys (wood, cork, and mold-injected plastic composition) have been used successfully for generations to attract waterfowl, reinforcing the importance of context, movement, and placement relative to the effigy's intended effect. Moreover, an effigy's decoy effect and its aversive effect can be one in the same. For example, investigative flight behaviors by some species in response to an effigy (e.g., woodpigeons [*Columba palumbus*]) might be ideal responses to the hunter, but they might ultimately avoid the effigy altogether (Murton et al. 1974, Inglis and Isaacson 1984).

Recent Advances

Avian response to object approach is critical in the contexts of predator detection, foraging, flocking, and avoiding collisions with static or moving structures (Martin 2011). As suggested above, there is potential to exploit sensory systems to enhance natural behavioral responses to object approach. Blackwell et al. (2009) examined responses to approach by a ground-based vehicle and vehicle-lighting regimen by

Fig. 2.2. Taxidermy mount of a turkey vulture suspended as an effigy to deter roosting by conspecifics. Photo credit: Thomas W. Seamans

brown-headed cowbirds (*Molothrus ater*) and mourning doves (*Zenaida macroura*), as well as properties of the visual system for both species. The authors found that vehicle lighting (i.e., the visual cue) can influence the avoidance behavior by cowbirds and that reaction to vehicle approach and light treatments was also affected by ambient light. Avoidance behavior by doves was not affected by lighting treatments, but doves became alert more quickly (on average by 3.3 s) than cowbirds. In contrast, cowbirds took flight sooner than doves. The authors also found that doves have a wider field of vision and can detect objects at a greater distance due to their higher visual acuity; however, cowbirds might flush earlier to reduce predation-risk costs associated with lower ability to visually track a given object. In extending their findings to reducing bird collisions with aircraft, Blackwell et al. (2009) suggested that there is potential to design vehicle-mounted

lighting that will enhance avian alert behavior and, subsequently, response to aircraft approach. However, the authors also recognized the role of species-specific antipredator strategies in response to approaching threats and that vehicle lighting might not yield the same behavioral responses across all bird species.

Some airports have incorporated use of radio-controlled (RC) aircraft to disperse birds (see Transport Canada 2002). As noted above, there is evidence that birds respond to full-size aircraft via antipredator behaviors (Bernhardt et al. 2010), and researchers now use RC aircraft to better understand how to exploit avian antipredator behaviors relative to aircraft approach (Blackwell et al. 2012*a*; S. Lima et al., Indiana State University, unpublished data). For instance, Blackwell et al. (2012*a*) fitted an RC aircraft with pulsing lights and calculated chromatic contrast (see above)

Fig. 2.3. Radio-controlled aircraft approaching a group of captive Canada geese in experiments designed to quantify detection of and response to aircraft approach and lighting treatment (Blackwell et al. 2012*a*). Photo credit: Gail Keirn

with lights on (pulsing) versus off. They estimated that Canada geese would perceive the aircraft with lights as a more visually salient object than the same aircraft without lights. The authors tested this hypothesis in a behavioral experiment measuring responses of geese to RC aircraft approaches (Fig. 2.3). They found that geese were alerted to the approach of the RC aircraft with the lights on 4 s earlier than with the lights off. Four seconds could be enough time for birds to engage in evasive maneuvers (Bernhardt et al. 2010) and to avoid a collision. Future studies will explore lights at other wavelengths based on the spectral sensitivity of the visual systems of bird species with a high frequency of strikes to enhance the observed behavioral response.

More recent RC aircraft designed to mimic raptors are proving effective in stimulating antipredator responses and dispersing flocking species, including gulls (E. Fernández-Juricic, unpublished data; see also Blackwell et al. 2012*a*).

Summary

Visual recognition of the treatment (e.g., postconsumption detection of a secondary repellent and the associated learned avoidance, stimulation of antipredator behaviors via predator effigies or laser dispersal, avoidance of disruptive stimuli such as fladry) is a common factor for nonlethal methods to deter wildlife from using areas or resources. In mammals, visual repellents generally rely on novelty or stimulation of antipredator behaviors. However, use of dogs to protect livestock might disrupt attacks by large predators (e.g., coyotes or wolves) but might not provoke a fear response (Gehring et al. 2010). Further, visual deterrents that rely on color detection by mammals must consider both visual capabilities of the target species and the context of application, for example, lighting cues (D'Angelo et al. 2006; Blackwell et al. 2002, 2012*a*; Blackwell and Seamans 2009). The context of application is critical with regard to birds, as well, but the complexities of avian visual configuration (Blackwell et al. 2009, Fernández-Juricic 2012) must also be understood. Specifically, is the visual deterrent or cue salient to the particular species in the given context? We contend that one can increase the period of effectiveness of a visual deterrent and decrease the degree of habituation by considering the sensory and behavioral ecology of the target species, the context of application, and how the method might be integrated with other techniques to enhance perception of predation risk (Ydenberg and Dill 1986, Frid and Dill 2002).

Future evaluations of visual deterrents used against wildlife, particularly in airport applications, should include integrating methods to enhance antipredator behavior. We encourage further investigation of the use of visual barriers (Blackwell et al. 2012b) against deer and exploitation of natural alarm signals in the form of effigies (Inglis and Isaacson 1984, Avery et al. 2002, Seamans 2004). In addition, we suggest that quantifying the effects of wavelength and pulse frequency of aircraft lighting, as well as chromatic and achromatic contrast of aircraft, can aid in enhancing avian response to aircraft approach (Blackwell et al. 2009, 2012a; Fernández-Juricic et al. 2011).

LITERATURE CITED

Altshuler, D. L. 2001. Ultraviolet reflectance in fruits, ambient light composition and fruit removal in a tropical forest. Evolutionary Ecology Research 3:767–778.

Avery, M. L., J. S. Humphrey, D. G. Decker, and A. P. McGrane. 1999. Seed color avoidance by captive red-winged blackbirds and boat-tailed grackles. Journal of Wildlife Management 63:1003–1008.

Avery, M. L., J. S. Humphrey, E. A. Tillman, K. O. Phares, and J. E. Hatcher. 2002. Dispersing vulture roosts on communication towers. Journal of Raptor Research 36:45–50.

Belant, J. L., P. P. Woronecki, R. A. Dolbeer, and T. W. Seamans. 1998. Ineffectiveness of five commercial deterrents for nesting starlings. Wildlife Society Bulletin 25:264–268.

Bernhardt, G. E., B. F. Blackwell, T. L. DeVault, and L. Kutschbach-Brohl. 2010. Fatal injuries to birds from collisions with aircraft reveal antipredator behaviors. Ibis 151:830–834.

Blackwell, B. F. 2002. Understanding avian vision: the key to using light in bird management. Proceedings of the Vertebrate Pest Conference 20:146–152.

Blackwell, B. F., and G. E. Bernhardt. 2004. Efficacy of aircraft landing lights in stimulating avoidance behavior in birds. Journal of Wildlife Management 68:725–732.

Blackwell, B. F., G. E. Bernhardt, and R. A. Dolbeer. 2002. Lasers as non-lethal avian repellents. Journal of Wildlife Management 66:250–258.

Blackwell, B. F., T. L. DeVault, T. W. Seamans, S. L. Lima, P. Baumhardt, and E. Fernández-Juricic. 2012a. Exploiting avian vision with aircraft lighting to reduce bird strikes. Journal of Applied Ecology 49:758–766.

Blackwell, B. F., E. Fernández-Juricic, T. W. Seamans, and T. Dolans. 2009. Avian visual configuration and behavioural response to object approach. Animal Behaviour 77:673–684.

Blackwell, B. F., and T. W. Seamans. 2009. Enhancing the perceived threat of vehicle approach to white-tailed deer. Journal of Wildlife Management 73:128–135.

Blackwell, B. F., T. W. Seamans, J. L. Belant, K. C. VerCauteren, and L. A. Tyson. 2012b. Exploiting antipredator behavior in white-tailed deer for resource protection. Wildlife Society Bulletin 36:546–553.

Bleich, V. C., R. T. Bowyer, A. M. Pauli, M. C. Nicholson, and R. W. Anthes. 1994. Mountain sheep, Ovis canadensis, and helicopter surveys: ramifications for the conservation of large mammals. Biological Conservation 70:1–7.

Blokpoel, H. 1976. Bird hazards to aircraft: problems and prevention of bird/aircraft collisions. Clarke, Irwin, Ottawa, Ontario, Canada.

Blumstein, D. T., J. C. Daniel, A. S. Griffin, and C. S. Evans. 2000. Insular tammar wallabies (Macropus eugenii) respond to visual but not acoustic cues from predators. Behavioral Ecology 11:528–535.

Boag, D. A., and V. Lewin. 1980. Effectiveness of three waterfowl deterrents on natural and polluted ponds. Journal of Wildlife Management 44:145–154.

Born, E. W., F. F. Riget, R. Dietz, and D. Andriashek. 1999. Escape responses of hauled out ringed seals (Phoca hispida) to aircraft disturbance. Polar Biology 21:171–178.

Caldwell, G. S., and R. W. Rubinoff. 1983. Avoidance of venomous sea snakes by naive herons and egrets. Auk 100:195–198.

Caro, T. 2005. Antipredator defenses in birds and mammals. Chicago University Press, Chicago, Illinois, USA.

Cleary, E. C., and R. A. Dolbeer. 2005. Wildlife hazard management at airports. Second edition. U.S. Department of Transportation, Federal Aviation Administration, Office of Airport Safety and Standards, Washington, D.C., USA.

Conover, M. R. 1982. Comparison of two behavioral techniques to reduce bird damage to blueberries: methiocarb and a hawk-kite predator model. Wildlife Society Bulletin 10:211–216.

Conover, M. R. 1985. Protecting vegetables from crows using an animated crow-killing owl model. Journal of Wildlife Management 49:643–645.

Conover, M. R. 1987. Acquisition of predator information by active and passive mobbers in ring-billed gull colonies. Behaviour 102:41–57.

Conover, M. R. 2002. Resolving human–wildlife conflicts. CRC Press, Boca Raton, Florida, USA.

Conover, M. R., and J. J. Perito. 1981. Response of starlings to distress calls and predator models holding conspecific prey. Zeitschrift für Tierpsychologie 57:163–172.

Cuthill, I. C. 2006. Color perception. Pages 3–40 in G. E. Hill and K. J. McGraw, editors. Bird coloration: mechanisms and measurements. Harvard University Press, Cambridge, Massachusetts, USA.

D'Angelo, G. J., J. G. D'Angelo, G. R. Gallagher, D. A. Osborn, K. V. Miller, and R. J. Warren. 2006. Evaluation of wildlife warning reflectors for altering white-tailed deer behavior along roadways. Wildlife Society Bulletin 34:1175–1183.

D'Angelo, G. J., A. Glasser, M. Wendt, G. A. Williams, D. A. Osborn, G. R. Gallagher, R. J. Warren, K. V. Miller, and M. T. Pardue. 2008. Visual specialization of an herbivore prey species, the white-tailed deer. Canadian Journal of Zoology 86:735–743.

de Almeida, A., H. T. Zarate do Couto, and A. F. de Almeida. 2010. Camouflaging of seeds treated with pesticides mitigates the mortality of wild birds in wheat and rice crops. Scientia Agricola 67:76–182.

Dolbeer, R. A., P. P. Woronecki, and L. Bruggers. 1986. Reflecting tapes repel blackbirds from millet, sunflowers, and sweet corn. Wildlife Society Bulletin 14:418–425.

Dowling, J. E. 2012. The retina: an approachable part of the brain. Belknap Press of Harvard University Press, Cambridge, Massachusetts, USA.

Dukes, T. W. 1969. The ocular fundus of normal white-tailed deer (Odocoileus virginianus). Bulletin of the Wildlife Disease Association 5:16–17.

Endler, J. A. 1990. On the measurement and classification of colour in studies of animal colour patterns. Biological Journal of the Linnean Society 41:315–352.

Endler, J. A. 1992. Signals, signal conditions, and the direction of evolution. American Naturalist 139:S125–S153.

Endler, J. A., and P. W. Mielke Jr. 2005. Comparing entire colour patterns as birds see them. Biological Journal of the Linnean Society 86:405–431.

Endler, J. A., and M. Théry. 1996. Interacting effects of lek placement, display behavior, ambient light, and color patterns in three neotropical forest-dwelling birds. American Naturalist 148:421–452.

Fernández-Juricic, E. 2012. Sensory basis of vigilance behavior in birds: synthesis and future prospects. Behavioural Processes 89:143–152.

Fernández-Juricic, E., J. Gaffney, B. F. Blackwell, and P. Baumhardt. 2011. Bird strikes and aircraft fuselage color: a correlational study. Human–Wildlife Interactions 5:224–234.

Finger, E., and D. Burkhardt. 1994. Biological aspects of bird coloration and avian color vision including ultraviolet range. Vision Research 34:1509–1514.

Frid, A., and L. M. Dill. 2002. Human-caused disturbance stimuli as a form of predation risk. Conservation Ecology 6:11. http://www.consecol.org/vol6/iss1/art11.

Gamberale-Stille, G., and T. Guilford. 2003. Contrast versus colour in aposematic signals. Animal Behaviour 65:1021–1026.

Gehring, T., K. C. VerCauteren, and J-M Landry. 2010. Livestock protection dogs in the 21st century: is an ancient tool relevant to modern conservation challenges? Bioscience 60:299–308.

Gomez, D. 2006. AVICOL: a program to analyze spectrometric data. https://sites.google.com/site/avicolprogram/.

Gorenzel, P. W., B. F. Blackwell, G. D. Simmons, T. P. Salmon, and R. A. Dolbeer. 2002. Evaluation of lasers to disperse American crows from night roosts. International Journal of Pest Management 48:327–331.

Goth, A. 2001. Innate predator-recognition in Australian brush-turkey (Alectura lathami, Megapodiidae) hatchlings. Behaviour 138:117–136.

Gouras, P. 2007. Webvision: the organization of the retina and visual system. http://webvision.med.utah.edu.

Guilford, T. 1990. The evolution of aposematism. Pages 23–61 in D. L. Evans and J. O. Schmidt, editors. Insect defenses: adaptive mechanisms and strategies of prey and predators. State University of New York, Albany, USA.

Guilford, T., and M. S. Dawkins. 1991. Receiver psychology and the evolution of animal signals. Animal Behaviour 41:1–14.

Hart, N. S. 2001. Variations in cone photoreceptor abundance and the visual ecology of birds. Journal of Comparative Physiology 187A:685–698.

Hart, N. S., and D. M. Hunt. 2007. Avian visual pigments: characteristics, spectral tuning, and evolution. American Naturalist 169:S7–S26.

Herrera, C. M. 1985. Aposematic insects as six-legged fruits: incidental short-circuiting of their defense by frugivorous birds. American Naturalist 126:286–293.

Honkavaara, J., H. Siitari, and J. Viitala. 2004. Fruit colour preferences of redwings (Turdus iliacus): experiments with hand-raised juveniles and wild-caught adults. Ethology 110:445–457.

Inglis, I. R., and A. J. Isaacson. 1984. The responses of woodpigeons (Columba palumbus) to pigeon decoys in various postures: a quest for a supernormal alarm stimulus. Behaviour 90:224–240.

Jacobs, G. H., J. F. Deegan II, M. A. Crognale, and J. A. Fenwick. 1993. Photopigments of dogs and foxes and their implications for canid vision. Visual Neuroscience 10:173–180.

Jacobs, G. H., J. F. Deegan II, J. Neitz, B. P. Murphy, K. V. Miller, and R. L. Marchinton. 1994. Electrophysiological measurements of spectral mechanisms in the retinas of two cervids: white-tailed deer (Odocoileus virginianus) and fallow deer (Dama dama). Journal of Comparative Physiology 174A:551–557.

Koehler, A. E., R. E. Marsh, and T. P. Salmon. 1990. Frightening methods and devises/stimuli to prevent mammal damage—a review. Proceedings of the Vertebrate Pest Conference 14:168–173.

Land, M. F., and D. E. Nilsson. 2002. Animal eyes. Oxford University Press, Oxford, United Kingdom.

Langley, W. M. 1983. Relative importance of the distance senses in grasshopper mouse predatory behavior. Animal Behaviour 31:199–205.

Lindström, L., R. V. Alatalo, and J. Mappes. 1999. Behavioral ecology: reactions of hand-reared and wild-caught predators toward warningly colored, gregarious, and conspicuous prey. Behavioral Ecology 10:317–322.

Martin, G. 2011. Understanding bird collisions with man-made objects: a sensory ecology approach. Ibis 153:239–254.

Marzluff, J. M., J. Walls, H. N. Cornell, J. C. Withey, and D. P. Craig. 2010. Lasting recognition of threatening people by wild American crows. Animal Behaviour 79:699–707.

Mason, J. R. 1998. Mammal repellents: options and considerations for development. Proceedings of the Vertebrate Pest Conference 18:325–329.

Mason, J. R., L. Clark, and N. J. Bean. 1993. White plastic flags repel snow geese (Chen caerulescens). Crop Protection 12:497–500.

McIlwain, J. T. 2006. An introduction to the biology of vision. Cambridge University Press, Cambridge, United Kingdom.

Montgomerie, R. 2006. Analyzing colors. Pages 90–147 in G. E. Hill and K. J. McGraw, editors. Bird coloration: mechanisms and measurements. Harvard University Press, Cambridge, Massachusetts, USA.

Murton, R. K. 1974. The use of biological methods in the control of vertebrate pests. Page 211 in D. Price-Jones and M. E. Solomon, editors. Biology in pest and disease control. Blackwell Scientific, Oxford, United Kingdom.

Murton, R. K., N. J. Westwood, and A. J. Isaacson. 1974. A study of woodpigeon shooting: the exploitation of a natural animal population. Journal of Applied Ecology 11:61–81.

Musiani, M., C. Mamo, L. Boitani, C. Callaghan, C. C. Gates, L. Mattei, E. Visalberghi, S. Breck, and G. Volpi. 2003. Wolf depredation trends and the use of fladry barriers to protect livestock in western North America. Conservation Biology 17:1538–1547.

Poulton, E. B. 1890. The colours of animals: their meaning and use, especially considered in the case of insects. Kegan Paul, Trench, Trübner.

Querubin, A., H. R. Lee, J. M. Provis, and K. M. B. O'Brien. 2009. Photoreceptor and ganglion cell topographies correlate with information convergence and high acuity regions in the adult pigeon (Columba livia) retina. Journal of Comparative Neurology 517:711–722.

Rondinini, C., and L. Boitani. 2007. Systematic conservation planning and the cost of tackling conservation conflicts with large carnivores in Italy. Conservation Biology 21:1455–1462.

Rubinoff, I., and C. Kropach. 1970. Differential reactions of Atlantic and Pacific predators to sea snakes. Nature 228:1288–1290.

Schnidrig-Petrig, R., and P. Ingold. 2001. Effects of paragliding on alpine chamois Rupicapra rupicapra rupicapra. Wildlife Biology 7:285–294.

Schuler, W. 1982. Zur function von warnfarben: die reaction junger stare auf wespenähnlich swartz-gelbe attrappen. Zeitschrift für Tierpsychologie 58:66–78.

Schuler, W., and E. Hesse. 1985. On the function of warning coloration: a black and yellow pattern inhibits prey-attack by naïve domestic chicks. Behavioral Ecology and Sociobiology 16:249–255.

Schwab, I. R. 2012. Evolution's witness: how eyes evolved. Oxford University Press, Oxford, United Kingdom.

Seamans, T. W. 2004. Response of roosting turkey vultures to a vulture effigy. Ohio Journal of Science 104:136–138.

Seamans, T. W., and G. E. Bernhardt. 2004. Response of Canada geese to dead goose effigy. Proceedings of the Vertebrate Pest Conference 21:104–106.

Sherman, D. E., and A. E. Barras. 2004. Efficacy of a laser device for hazing Canada geese from urban areas of northeast Ohio. Ohio Journal of Science 104:38–42.

Shivik, J. A., A. Treves, and P. Callahan. 2003. Nonlethal techniques for managing predation: primary and secondary repellents. Conservation Biology 17:1531–1537.

Sillman, A. 1973. Avian vision. Pages 349–387 in D. S. Farner, J. R. King, and K. C. Parkes, editors. Avian biology. Volume 3. Academic Press, New York, New York, USA.

Smith, S. M. 1975. Innate recognition of coral snake pattern by a possible avian predator. Science 187:759–760.

Smith, S. M. 1977. Coral–snake pattern recognition and stimulus generalization by naïve Great Kiskadees (Aves: Tyrannidae). Nature 265:535–536.

Sodhi, N. 2002. Competition in the air: birds versus aircraft. Auk 119:587–595.

Stankowich, T., and D. T. Blumstein. 2005. Fear in animals: a meta-analysis and review of risk assessment. Proceedings of the Royal Society B 272:2627–2634.

Tinbergen, N. 1948. Social releasers and the experimental method required for their study. Wilson Bulletin 60:6–51.

Transport Canada. 2002. Wildlife control procedures manual. TP 11500 E. Third edition. Transport Canada, Safety and Security, Aerodrome Safety Branch, Ottawa, Ontario, Canada.

Veen, T., D. S. Richardson, K. Blaakmeer, and J. Komdeur. 2000. Experimental evidence for innate predator recognition in the Seychelles warbler. Proceedings of the Royal Society B 267:2253–2258.

VerCauteren, K. C., J. M. Gilsdorf, S. E. Hygnstrom, P. Fioranelli, J. A. Wilson, and S. Barras. 2006. Green and blue lasers are ineffective for dispersing deer at night. Wildlife Society Bulletin 34:371–374.

VerCauteren, K. C., and M. J. Pipas. 2003. A review of color vision in white-tailed deer. Wildlife Society Bulletin 31:684–691.

Vorobyev, M., and D. Osorio. 1998. Receptor noise as a determinant of colour thresholds. Proceedings of the Royal Society B 265:351–358.

Vorobyev, M., D. Osorio, A. T. D. Bennett, N. J. Marshall, and I. C. Cuthill. 1998. Tetrachromacy, oil droplets and bird plumage colours. Journal of Comparative Physiology 183A:621–633.

Walls, G. L. 1942. The vertebrate eye and its adaptive radiation. Hafner, New York, New York, USA.

Warren, R. J., K. V. Miller, and M. T. Pardue. 2008. Visual specialization of an herbivore prey species, the white-tailed deer. Canadian Journal of Zoology 86:735–743.

Windberg, L. A. 1997. Coyote responses to visual and olfactory stimuli related to familiarity with an area. Canadian Journal of Zoology 74:2248–2253.

Ydenberg, R. C., and L. M. Dill. 1986. The economics of fleeing from predators. Advances in the Study of Behavior 16:229–249.

3

Larry Clark
Michael L. Avery

Effectiveness of Chemical Repellents in Managing Birds at Airports

Repellents include methods and devices used to manipulate behavior of animals to reduce damage or nuisance. Critical to the design and success of repellents is understanding how sensory modalities mediate perception of signals, and how ecological context and sensory inputs influence animal learning. A repellent's success is tied to the axiom of using the proper tool for the proper job. When repellents "fail," it is almost always because wildlife managers have not appropriately matched signal, receiving systems, message, and context. Reconciling such considerations can be a complex process. In this chapter we review components and processes essential for the successful use of repellents for managing birds at airports.

There is often great disappointment among managers on the performance of repellents, and chemical repellents may be among the most misunderstood wildlife management tools. Perceived failures of chemical repellents are not always accurate, as performance is aligned with the sensory biology of the target animal and context of application. Successful use of repellents requires (1) understanding the rules of animal learning; (2) understanding the sensory abilities of targeted animals; (3) appreciating that repellents are tools to shape and modify behavior, not toxicants; (4) understanding that repellents train individual animals and that, when populations turn over frequently, constant training may be required; and (5) understanding that repellents work best if alternative resources or places are available, and that if alternatives are not available, the animal may endure unpleasant side effects. In short, large numbers of animals with no alternative resources or places to go will swamp the defensive characteristics of a repellent. Given use based on the requisites described above, repellents can be effective and deserve a place in integrated and adaptive pest management strategies. The reviews on these topics should be sought for in-depth coverage (Garcia and Hankins 1977; Revusky 1977; Dooling 1982; Kare and Brand 1986; Clark 1997, 1998b; Mason and Clark 1997, 2000; Reidinger 1997; Domjan 1998; Conover 2002; Werner and Clark 2003).

Mediating Sensory Modalities

Mediating sensory modalities in birds includes the chemical senses (smell or olfaction, taste or gustation, irritation), hearing or audition, vision, and touch (see also Chapters 2 and 4). In general, birds have excellent auditory and visual capabilities and moderately developed chemical senses (Mason and Clark 2000, Walsh and Milner 2011).

Olfaction acts as a telereceptive system capable of receiving airborne chemical stimuli in extreme dilution over relatively great distances. Olfactory cues may generally orient some bird species toward food sources (e.g., Stager 1964, Verheyden and Jouventin 1994) or elicit specific discrimination behaviors from others (e.g., Clark and Mason 1987, Roper 1999). Gustation requires more intimate contact between the source of the signal and the receptors. Taste receptors in birds are located throughout the oral and pharyngeal cavities, and generally mediate sensory qualities also perceived by mammals: sweetness, saltiness, sourness, and bitter-

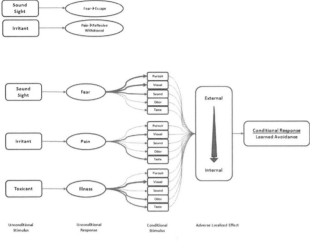

Fig. 3.1. Nonlearned and learned responses to stimuli for birds. The top illustration shows the likely pairings of sensory input and nonlearned behavioral responses. The bottom illustration shows the pairings between the nature of the unconditional stimulus and the effect it has on the animal (unconditional response), as well as the likelihood those pairings can be matched with an animal's mediating sensory modality (conditional stimulus). Thicker arrows indicate associations that are more likely. The more the unconditional stimulus (US)/unconditional response–conditional stimulus (CS) pairing is internalized in an animal, the stronger the CS-conditional response (CR) association is likely to be. The more likely the US/CR-CS pairing is self-limited by the animal, as might be the case for externally exposed cues, the weaker the CS-CR association is likely to be.

ness (Kare and Brand 1986). Bird species perceive taste qualities differently, however, reflecting their species-specific ecologies and food habits (Berkhoudt 1985). Chemesthesis is the perception of chemically irritating or painful stimuli. Noxious chemical stimuli may give rise to qualities such as stabbing, throbbing, burning, or itching, depending on the specific nociceptive fiber that is stimulated and the neurotransmitter released (Clark 1998a). Vision, like olfaction, is a telereceptive system (Zeigler and Bischof 1993). Unlike olfaction, the source of visual cues is more readily identified because of the linear relationship between source and receptor. Visual cues facilitate navigation, recognition of conspecifics and mates, predator avoidance, and food selection (Chapter 2). Sounds provide birds with information regarding territorial defense, mate selection, navigation, and recognition of predators, conspecifics, and prey location (Gill 1990, Beason 2004; Chapter 4). The sense of touch is integral to the feeding behavior of many birds, particularly many waders and shorebirds (Gill 1990; Chapter 4). Furthermore, while birds are flying, feathers are constantly being adjusted in response to tactile sensations received via nerves at the base of the shafts.

Types of Chemical Repellents

There are two fundamental repellent classes: primary and secondary (Clark 1997). A repellent is classified as primary or secondary based on the physiological mode of action and whether avoidance behavior is learned. Primary repellents possess a quality (e.g., unpalatable

taste, odor, irritation) that evokes reflexive withdrawal or escape behavior. Secondary repellents evoke an adverse physiological effect (e.g., illness), which the animal associates with a sensory cue (e.g., taste, odor, visual cue) and then learns to avoid. These definitions help to quickly assess the likely effectiveness of a chemical repellent in a particular ecological context.

Behavioral Bases for Repellency

Repellents evoke reflexive withdrawal, escape, or avoidance behaviors (Fig. 3.1). The behaviors differ, even though the manager may not be able to distinguish the underlying cause (characterized below) for an animal staying away from an object or area.

Reflexive Withdrawal

Painful or irritating stimuli may evoke innate, reflexive withdrawal from the stimulus. The response is adaptive because it precludes further damage or harm to the animal. In the parlance of learning psychology, the aversive stimulus is the unconditional stimulus (US) and the reflexive withdrawal is the unconditional response (UR). Because an animal limits exposure to potentially harmful stimuli, the degree and magnitude of exposure to the US are weak, and animals do not efficiently form learned associations to these types of stimuli (Clark 1996). Animals are therefore more apt to revisit sites or sample foods where the US produces a weak or external localized effect (see below).

Escape

Neophobia is generally associated with escape from a novel stimulus. Although escape behavior may confer a short-term advantage to an animal (i.e., the animal reduces its risk to the unknown), reliance on this behavioral paradigm is not a sound strategy for the animal in the long term, nor is it a sound management strategy. Habituation to the US may occur when there is no negative reinforcement. From the manager's perspective, habituation is the unwanted, learned response that the stimulus has no consequence. A classic example is the use of owl effigies to repel birds from an area (Chapter 2). A predator that does not pursue its prey quickly loses its perception as a threat.

Avoidance

Avoidance behaviors involve learning. The manager uses a repellent to train target animals to avoid an object or place. There are four elements in the formation of a learned avoidance response (Fig. 3.1). The repellent (US) elicits an unpleasant experience (UR) for the animal. The animal associates the UR with sensory cues (conditional stimulus, or CS) paired in space and time to form the learned avoidance (conditional response, or CR; Pavlov 1906, Garcia et al. 1966).

A widely known strategy for learned avoidance is the conditioned flavor avoidance paradigm (CFA), also known as conditioned taste aversion (CTA; Garcia et al. 1966, Garcia 1989). The former is a better characterization because it acknowledges the multi-sensory nature of oral exposure to stimuli (inclusive of taste, retronasal olfaction, chemical irritation, tactile cues). However, myriad UR-CS pairings exist, and some associations are more frequently paired in nature and hence are more readily established (Milgram et al. 1977). Most mammals readily develop aversions based on flavor cues (taste, odor, irritation) and gastrointestinal illnesses (Revusky 1977). In contrast, birds are less apt to form CFAs, whereas they are more likely to form aversions based on visual cues and gastrointestinal illness (Mason and Reidinger 1983; Fig. 3.1). The likelihood and strength of learned aversions based on sensory inputs have a neurophysiological basis that differs among taxa (Provenza 1995). For this

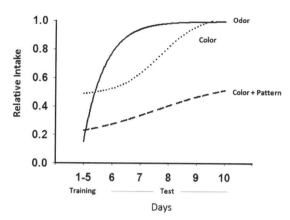

Fig. 3.2. Extinction curves for learned avoidance when unconditional stimuli are irritants in European starlings. Conditional stimuli (CS) were odor (solid line), colored target (dotted line), and colored-patterned target (dashed line). The curves indicate that visual cues are better CS in birds than are chemical cues, and that synergy can be observed for visual cues. Data derived from Mason (1989) and Clark (1996)

reason, development and application of effective repellents (i.e., reliable CRs) depend on understanding how an animal will sense and integrate the adverse experience.

Salience

Evolution (i.e., the development of sensory systems and neurophysiological interactions of sensory afferents) and ecological context constrain the salience of cues. In laboratory tests, European starlings (*Sturnus vulgaris*) were offered food treated with a chemical irritant paired with a sensory cue (either an odor, colored target, or colored target and pattern) during a five-day training period (Mason 1989, Clark 1996). Over the following five days, the starlings received unadulterated food paired with the odor or visual targets. Starlings avoided the color and pattern targets longer than the colored target alone, and exhibited almost no long-term avoidance when the sensory cue was an odor (Fig. 3.2).

The experiments demonstrate two things. First, starlings were more attuned to visual cues as conditional stimuli. Second, adding unique, independent features to the visual cue enhanced learned avoidance, both in magnitude and duration. These results are to be expected for species such as birds that have a well-developed visual sensory system for color and pattern (Endler and Théry 1996). For most mammals, a similar experiment would place pattern and odor above color as the most salient sensory cues, because most mammals have poorly developed color vision but highly developed abilities to detect edges and motion (Jacobs 2009; Chapter 2).

Adverse Localized Effect

The intensity and duration of a learned avoidance response depend on the degree and location of the aversive experience (UR). The learned response is weakest when the animal has control over its exposure—when the animal can escape or withdraw from peripherally applied repellents (sound, sight, or chemical if delivered to mouth, eyes, or skin). The learned avoidance is strongest when the animal cannot escape the application of the US (e.g., a chemical that produces a gastrointestinal illness; Pelchat et al. 1983).

Concurrent Interference

Specific pairings of stimuli can influence what an animal learns, and understanding how can help a manager deploy effective repellent methods. An example begins with a bear (Ursidae) visiting a dump because it positively associates the dump with food. The US is food, the UR is caloric reward/satiety, the CS is food odor (or some other sensory aspect of dumps), and the CR is dump visitation. A ranger shoots the bear with rubber bullets in an attempt to train it to avoid the dump. From the park ranger's perspective, the rubber bullet is the US and the unpleasant experience—pain—is the UR. The ranger believes that the bear will associate the UR with the CS (i.e., connect the dump's visual cues or odors with pain, meaning the bear will avoid the dump). But the bear continues to visit the dump, only fleeing when the park ranger shows up. What went wrong? Concurrent interference (Sayre and Clark 2001), or the presentation of competing cues that selectively attend to the most relevant cues in an animal's learned associations between cause and effect (Fig. 3.3). Normally the bear would be trained to avoid the sensory cues associated with the dump. However, the arrival of the park ranger was exactly correlated with the pain from the

Scenario 1: Animal learns C_1 is the harmful agent, C_3 & C_2 are safe.

$C_1C_2 \rightarrow AE_{100\%}$

$C_3C_2 \rightarrow AE_{0\%}$

Scenario 2: Animal learns C_2 is the harmful agent, C_3 & C_1 are safe.

$C_1C_2 \rightarrow AE_{50\%}$

$C_3C_2 \rightarrow AE_{5\%}$

Fig. 3.3. Two scenarios of how pairings of cues influence inference about safety and harmfulness of a cue. C_1 is the sensory cue presented to an animal, AE is the aversive experience, and its subscript percentage is the frequency with which that aversive experience occurs.

rubber bullet, representing a higher degree of salience to the bear because the US was only experienced at the dump when the ranger was present. Had the ranger taken precautions to be less obvious, the negative reinforcement of the rubber bullet would not have been so predictable, and the bear might have learned to avoid the dump.

Ecological Context

Ideally, a repellent moves animals from an undesired place to a place where their presence is accepted. After successful application of a bird or mammal repellent, the total number of animals will not decrease, but they will be distributed differently. A realistic goal of repellent application is therefore not to eliminate birds or mammals at a location but to reduce their numbers to an acceptable, manageable level. To the extent that a repellent can help redistribute the local wildlife population from sensitive areas to nonsensitive ones, it will be a successful component of an airport wildlife management plan.

Understanding why wildlife species are attracted to a given site in the first place is central to determining the most effective strategy for moving them. Feeding opportunities are the most likely reasons for the presence of most animals (Chapter 8). Other possible explanations include nesting, roosting, access to drinking water, and refuge from predators. Making the resources unavailable to the animal eliminates its reason to be there. Whatever the resource, if it can be removed, the animals will no longer frequent the site. If physical removal of the resource is not possible, then the resource can sometimes be rendered unappealing or undesirable by application of a chemical repellent.

Birds have high metabolic rates and are constantly seeking readily accessible sources of food to meet their nutritional requirements with low expenditure of effort. This is especially important to young birds that are not experienced foragers. In the late summer and fall, newly fledged birds constitute a large portion of many foraging flocks. At other times of year, alternative sources of food may be limited or lacking altogether. Given this situation, it is easy to appreciate why wetlands and other resources at airports can be powerful attractions to animals. With substantial potential benefits to animals from using airports, there must be commensurately high potential costs in order to discourage them.

To be effective, a chemical repellent must alter the balance in the airport environment, either by greatly reducing the benefits of feeding or by greatly increasing the costs. Increasing the cost to the animals usually means increasing the amount of time and energy required to feed at that site. The more time the animal has to spend acquiring the requisite nutritional resources, the less time it can spend on other essential activities such as territorial defense, mate acquisition, provisioning young, body maintenance, predator vigilance, and so on. There is therefore substantial pressure on an animal to feed efficiently. Caloric gain is not the only nutritional requirement, but it seems pervasive. If it becomes difficult for the animal to maintain a certain rate of energy intake, foraging theory predicts that the animal will look for other sources of food (MacArthur and Pianka 1966). The net effect of applying a chemical repellent is to lower the value of the food source by reducing the animal's rate of energy intake. Incorporation of other methods such as pyrotechnics, lasers, or shooting might reinforce the effectiveness of the repellent (Chapters 2 and 4).

Primary chemical bird repellents do not generally promote strong learned avoidance responses (Clark 1996), as illustrated by field observations where a formulation of methyl anthranilate (Nachtman et al. 2000), a primary repellent, was incorporated into day-covering material and sprayed onto the open

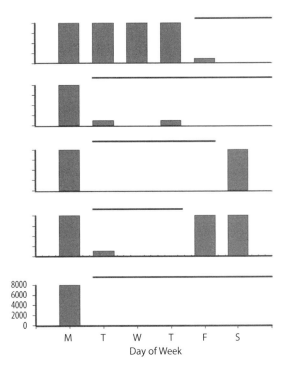

Fig. 3.4. Efficacy of a primary repellent at the Tullytown Landfill near Philadelphia, Pennsylvania, USA. Surface sprays containing 1% methyl anthranilate and yucca were applied several times per day (horizontal lines) throughout a five-week period in August 1994. Bar heights represent gull counts at the tipping face of the landfill and site where surface sprays were applied. Note that the formulation successfully repelled gulls from the site during intensive application. Gulls were always in close proximity, however. Sampling allowed the gulls to return as soon as the application of repellent was halted. Data source: L. Clark, unpublished data

tipping surface of Tullytown Landfill near Philadelphia, Pennsylvania, USA (Fig. 3.4). Gull (Laridae) counts were significantly lower after spraying. As long as spray coatings were applied, the gulls stayed away from the site. When spraying operations were stopped (e.g., weekends and holidays), however, gulls returned to the tipping site within 24 hr. The data are consistent with the interpretation that the repellent was effective at preventing the use of a valued resource because of its intrinsic irritating qualities, but did not have a paired salient cue that would promote long-term avoidance of the site. Does this mean that primary repellents are not useful? The answer is no. Depending on ecological context, the same repellent may be quite effective. At Tullytown Landfill, gulls had an alternative nearby resource: the untreated Groves Landfill (about 1 km away) and nearby roosting sites.

A similar field observational study was conducted at Dane County Landfill in Madison, Wisconsin, USA, with markedly different results (Fig. 3.5). A single application of the methyl anthranilate formulation was applied, and gulls left the site. The gulls did not begin to return until 21 days later. Over the course of the next week, gull numbers increased and a second application of repellent was applied. Again, gulls left the site and did not return over the course of the next ten days. Why did the Dane County site produce such different results from the Tullytown site? The difference at Dane County was that the gulls' roosting site was at least 25 km distant, and there were no nearby alternative foraging sites. These results are consistent with the ecological foraging concept known as central place foraging (MacArthur and Pianka 1966). Central place foraging occurs when birds travel from a roost or nesting site to peripheral feeding locations and return to the roost or nest site each night. The choice of foraging sites is presumed to be an optimization among effort, distance traveled, and reward (Stephens et al. 2007). In this case the successful use of the primary repellent is entirely dependent upon the knowledge of its mode of action, how that mode influences learned avoidance, and the ecological context under which the repellent is applied. At Dane County Landfill, the repellent was the proper tool for the job, but at Tullytown Landfill, it was not.

Reducing the value of the food source is a key component to repellent use. The other crucial factor is the availability of alternative sources of food. An animal with no alternatives will tolerate much greater discomfort than will one with access to other food sources. Thus chemical repellents function more effectively with an available selection of food sources than with no alternative. The disparity in attractiveness between the airport site and potential alternative feeding sites will influence how noxious the repellent must be to effect a change in the animal's behavior. In an airport context, availability of attractive alternate food will be somewhat challenging, as such sites must be sufficiently distant from the operations area so as not to create hazardous situations themselves. Establishment

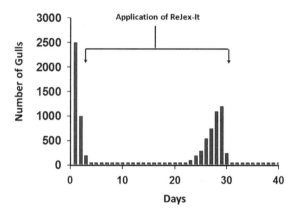

Fig. 3.5. Efficacy of a primary repellent at the Dane County Landfill, Madison, Wisconsin, USA. Two surface sprays containing a landfill coating of 1% methyl anthranilate and yucca were applied (arrows). Gulls left the landfill after one application, eventually returning during the third week after initial application. A second application reduced bird numbers at the site once more (L. Clark, unpublished data). The greater distance of available roost sites prevented frequent sample visitation by gulls; once repelled, the gulls tended to stay away, even though the repellent application was intermittent. This example illustrates how central place foraging can influence the efficacy of a repellent.

of feeding sites specifically to attract animals away from an airport is probably not intuitively pleasing, and the effectiveness of this management approach should be tested experimentally.

Water is a major attractant for wildlife, and airports usually include retention ponds and other permanent and ephemeral bodies of water (Chapter 9). Birds use such resources as feeding sites, and other wildlife are attracted because of loafing, bathing, and drinking opportunities. Consistent use of a fogger to disperse an aerosolized avian irritant such as methyl anthranilate along the periphery of the impoundment can likely change avian use patterns.

Chemical repellents are often applied to turf to repel Canada geese (*Branta canadensis*). The repellents are ingested while grazing, and the animal experiences pain (methyl anthranilate formulations) or a postingestional malaise (anthraquinone formulations; Dolbeer et al. 1998, Blackwell et al. 1999). In both cases the learned response is for geese to avoid feeding on the turf. When using turf repellents for geese, however, managers should be aware of possible underlying fac-

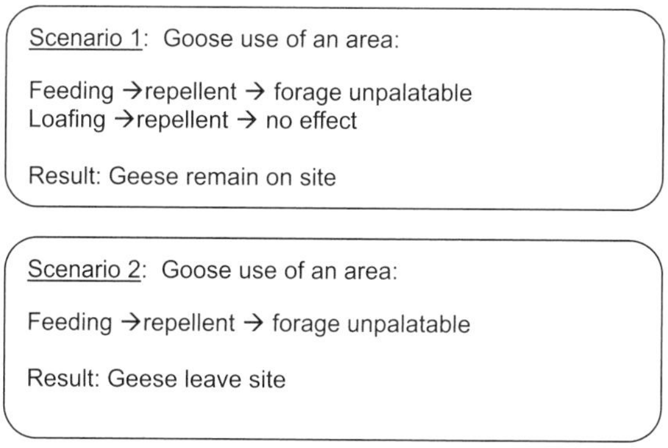

Scenario 1: Goose use of an area:

Feeding →repellent → forage unpalatable
Loafing →repellent → no effect

Result: Geese remain on site

Scenario 2: Goose use of an area:

Feeding →repellent → forage unpalatable

Result: Geese leave site

Fig. 3.6. Scenarios of how expectation can misinform a manager about a repellent's success. If Canada geese are using a site for feeding, then applying a repellent will render the forage unpalatable, and the geese will move. If geese are attracted to the site for loafing and feeding, however, then even though the repellent makes the food unpalatable, geese will continue loafing at the site.

tors that may motivate geese to use the site. Managers often believe the repellent has failed because geese stay in the treated area. Even if geese remain in the area, the repellent worked as designed: it stopped the feeding behavior of geese on the treated turf. The repellent is not designed to repel geese from an area. The geese may no longer graze, but the area may still be suitable for loafing. If the geese do leave the area, it is likely that the area was used only for foraging. Once forage is removed or unpalatable, the geese move on. The manager may misinterpret this as a successful application of the repellent in that geese stayed away from the area. But what is missing is an accurate assessment of why the area is being avoided (Fig. 3.6). Without such assessment, the manager may experience success on some occasions and failure on others. The manager may consequently abandon a perfectly good tool, thinking it is not consistently effective.

Migratory behaviors of many bird populations and seasonal availability of food resources combine to produce variability in numbers and types of birds attracted to a given facility. Some species of migrant and wintering birds are attracted to stands of wax myrtle (*Myrica cerifera*) because of seasonal availability of the waxy, lipid-rich berries (Place and Stiles 1992). Similarly, migrating barn swallows (*Hirundo rustica*) and tree swallows (*Tachycineta bicolor*) can descend in large flocks to exploit seasonally abundant swarms of insects. At air-

ports, timely application of aerosolized chemical bird repellents (Engeman et al. 2002) has provided relief from large aggregations of such birds.

Bird roosts at or near airport facilities can pose serious problems for airport managers, because birds often exploit food resources at these sites. Birds arriving and departing the roost can elevate the risk to low-flying aircraft, even without the birds being on airport property. Roosting aggregations of vultures (*Cathartes aura* and *Coragyps atratus*), crows (*Corvus brachyrhynchos*), or other birds might be successfully dispersed with a fogged or aerosolized repellent. Appropriate visual deterrents or effigies (Chapter 2), reinforced as needed with laser or pyrotechnic harassment, can usually disperse roosts (Avery et al. 2002, 2008; Teague 2002).

Landfills are often located near airports and represent a major food resource for many birds. Safe operation of an airport might therefore also require bird management at a landfill. Persistent harassment with pyrotechnics and lethal control using shotguns are standard bird control methods at landfills, and it is not clear if repellent applications can play a significant role, especially on a large scale. To reduce bird use of ponds or temporary wetlands, fogging with methyl anthranilate could potentially be effective (Belant et al. 1995). Because the working face of a landfill is continually turning over and because of constant heavy equipment traffic, effective repellent use would be difficult and possibly cost-prohibitive.

Deer and coyotes, attracted to food resources near airports, are the mammals most often involved in damaging aircraft collisions (e.g., Dolbeer et al. 2010, Biondi et al. 2011). Some chemical repellents can reduce browsing damage by deer to crops and ornamental plantings. Application of such repellents at airports is conceivable, providing deer are attracted to discrete, identifiable food sources that can be readily treated with a repellent and that cannot be managed in other, more permanent ways. In addition, predator urine can potentially inhibit deer use of a given area (Swihart et al. 1991, Nolte et al. 1994), although this application in airport situations is untested. For coyotes, chemical irritants and aversive agents have been tested and evaluated, mostly for livestock protection (Mason et al. 2001, Shivik 2004). To date, there is no indication that any chemical repellent method tested will by itself repel coyotes from airports.

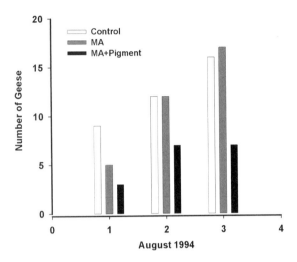

Fig. 3.7. Synergistic effects of sensory stimuli, an irritating repellent, methyl anthranilate, and a visual cue. When pigment (titanium dioxide) alone was applied to turf, geese grazed on the plots, and the use of those plots increased over time (an example of neophobia and eventual habituation to the novel stimulus; white bars). The bird irritant, methyl anthranilate (1%), had a more pronounced effect in the first week, but eventually the geese used those plots with increasing frequency (gray bars). The combination of the two cues (black bars), however, each of which yielded suboptimal avoidance, proved to be a highly effective deterrent. Data source: L. Clark, unpublished data

Integrated Management

If the target animal population is resident in the area of the airport facility, then a learned aversive response might be preferable, and a secondary repellent such as anthraquinone can be an effective management tool (Avery et al. 1998, Werner et al. 2011). But if the target population is transient, such as a wave of migratory birds, then conditioning individuals is not feasible. Instead, a more effective management approach is application of a primary repellent, such as methyl anthranilate, that produces an unlearned aversive reaction.

Chemical repellents, along with other wildlife damage management devices such as visual or aural deterrents, can expand the collective impact of management methods beyond their individual impact (Fig. 3.7). The appropriate combination or integration of methods and techniques is dynamic, contingent on local circumstances and the specifics of the pest animal population. Conditions at a given facility change seasonally at least,

so the management prescription should change accordingly. Regardless of target species, any value in the application of aversive chemical stimuli will come from integrated management approaches, including lethal and nonlethal techniques. For successful outcomes, ingenuity and resourcefulness are called for. Both anthraquinone and methyl anthranilate are registered as avian grazing deterrents on turf. Because they have different modes of action, their use in combination might provide greater impact than either used alone. Staggering the use of different repellents temporally or creating simultaneous patches with different repellents across sensitive areas on the airport facility will reduce the likelihood of habituation and will challenge the animal's perception of the local environment as it seeks acceptable foraging sites.

Available Repellents

The number of chemical repellents available for wildlife management is limited. Currently there are only two registered chemical bird repellents in the USA: those that contain methyl anthranilate and those that contain anthraquinone. Birds do not perceive capsaicin as irritating (Clark 1998a), and claims to the contrary are simply not scientifically credible. A number of chemicals have been tested as primary and secondary bird repellents (Clark 1997); however, most of these compounds are not available via U.S. Environmental Protection Agency registrations. Table 3.1 lists the products available for legal use in the USA (in addition to those containing methyl anthranilate and anthraquinone).

Summary

The effective use of chemical repellents to deter wildlife from airport environments requires an understanding of how animals learn from negative experiences as well as the sensory abilities of the target species. This information is critical in discerning the appropriate repellent for a particular behavioral context. In addition, alternative resources should be available and chemical repellents should be integrated with other management methods. Finally, use of chemical repellents must follow the guidelines set by the specific U.S. Environmental Protection Agency registration.

Table 3.1. Products and active ingredients of bird and mammal repellents registered by the U.S. Environmental Protection Agency (EPA). Label restrictions may apply.

EPA regulation no.	Active ingredient	Product	Species
	Agriculture crops, ornamentals		
070703	Red pepper		Coyotes, dogs
100628	Meat meal		Rabbits, raccoons, deer
	Landscape, yards		
66676-1-23	Denatonium benzoate	Ortho deer repellent	Deer
270-233	Morpholine, thiram	Detour deer and rabbit repellent	Deer, rabbits
64864-26	Ammonia soaps of higher fatty acids	Hinder deer and rabbit repellent	Deer, rabbits
67356-1-4	Whole egg solids	Bonide deer and rabbit repellent	Deer, rabbits
67356-2-4	Capsaicin, garlic extract	Bonide deer and rabbit repellent	Deer, rabbits
779-29-56644	Tobacco, naphthalene	Repel pet and stray repellent	Dogs, cats
59578-2-4626	Methyl nonyl ketone	XP-20 dog and cat repellent	Dogs, cats
779-29-4	Nicotine, naphthalene, animal blood, denatured	Bonide shotgun dog and rabbit repellent	Dogs, rabbits
64439-1-4	Castor oil	Bonide shotgun mole repellent	Moles, gophers
4-403	Ziram	Bonide rabbit scat	Rabbits
45735-2	Thymol: benzyldiethyl (2,6,xylyl carbamoyl) methyl ammonium sacchari	Ropel animal, rodent, and bird repellent	Rodents, birds
	Ornamentals		
122401	Fish oil		Rabbits, deer
	Ornamentals, nonfood		
125001	1-butanethiol		Deer
	Pet care		
11715-13-270	Methyl nonyl ketone, petroleum gases, liquefied, sweetened	Repel II dog and cat repellent	Dogs, cats
45987-1-270	Dihydro-5-pentyl-2(3H)-furanone 2(3H)-furanone, 5-heptyldihydro d-limonene	Repel II dog and cat repellent	Dogs, cats

LITERATURE CITED

Avery, M. L., J. S. Humphrey, T. M. Primus, D. G. Decker, and A. P. McGrane. 1998. Anthraquinone protects rice seed from birds. Crop Protection 17:225–230.

Avery, M. L., J. S. Humphrey, E. A. Tillman, K. O. Phares, and J. E. Hatcher. 2002. Dispersal of vulture roosts on communication towers. Journal of Raptor Research 36:44–49.

Avery, M. L., E. A. Tillman, and J. S. Humphrey. 2008. Effigies for dispersing urban crow roosts. Proceedings of the Vertebrate Pest Conference 23:84–87.

Beason, R. C. 2004. What can birds hear? Proceedings of the Vertebrate Pest Conference 21:92–96.

Belant, J. L., S. W. Gabrey, R. A. Dolbeer, and T. W. Seamans. 1995. Methyl anthranilate formulations repel gulls and mallards from water. Crop Protection 14:171–175.

Berkhoudt, H. 1985. Structure and function of avian taste receptors. Pages 463–495 in A. S. King and J. McLelland, editors. Form and function in birds. Volume 3. Academic Press, London, United Kingdom.

Biondi, K. M., J. L. Belant, J. A. Martin, T. L. DeVault, and G. Wang. 2011. White-tailed deer incidents with U.S. civil aircraft. Wildlife Society Bulletin 35:303–309.

Blackwell, B. F., T. W. Seamans, and R. A. Dolbeer. 1999. Plant growth regulator (Stronghold™) enhances repellency of anthraquinone formulation (Flight Control™) to Canada geese. Journal of Wildlife Management 63:1336–1343.

Clark, L. 1996. Trigeminal repellents do not promote conditioned odor avoidance in European Starlings. Wilson Bulletin 108:36–52.

Clark, L. 1997. A review of 117 carbocyclic compounds. Pages 343–352 in J. R. Mason, editor. Repellents in wildlife management. National Wildlife Research Center, Fort Collins, Colorado, USA.

Clark, L. 1998a. Physiological, ecological, and evolutionary bases for the avoidance of chemical irritants by birds. Current Ornithology 14:1–37.

Clark, L. 1998b. Review of bird repellents. Proceedings of the Vertebrate Pest Conference 18:330–337.

Clark, L., and J. R. Mason. 1987. Olfactory discrimination of plant volatiles by the European starling. Animal Behaviour 35:227–235.

Conover, M. R. 2002. Resolving human–wildlife conflicts: the science of wildlife damage management. CRC Press, Boca Raton, Florida, USA.

Dolbeer, R. A., T. W. Seamans, B. F. Blackwell, and J. L. Belant. 1998. Anthraquinone formulation (Flight Control™) shows promise as avian feeding repellent. Journal of Wildlife Management 62:1558–1564.

Dolbeer, R. A., S. E. Wright, J. Weller, and M. J. Begier. 2010. Wildlife strikes to civil aircraft in the United States 1990–2009. Serial Report 16. U.S. Department of Transportation, Federal Aviation Administration, Office of Airport Safety and Standards, Washington, D.C., USA.

Domjan, M. 1998. The principles of learning and behavior. Fourth edition. Brooks/Cole, Pacific Grove, California, USA.

Dooling, R. J. 1982. Auditory perception in birds. Pages 95–130 in D. E. Kroodsma and E. H. Miller, editors. Acoustic communication in birds. Academic Press, New York, New York, USA.

Endler, J. A., and M. Théry. 1996. Interacting effects of lek placement, display behavior, ambient light, and color patterns in three neotropical forest-dwelling birds. American Naturalist 148:421–452.

Engeman, R. M., J. Peterla, and B. Constantin, 2002. Methyl anthranilate aerosol for dispersing birds from the flight lines at Homestead Air Reserve Station. International Biodeterioration and Biodegradation 49:175–178.

Garcia, J. 1989. Food for Toman: cognition and cathexis in concert. Pages 45–85 in T. Archer and L. Nilson, editors. Aversion, avoidance and anxiety. Erlbaum, Hillsdale, New Jersey, USA.

Garcia, J., and W. G. Hankins. 1977. On the origin of food aversion paradigms. Pages 3–19 in L. Baker, M. Domjan, and M. Best, editors. Learning mechanisms in food selection. Baylor University Press, Waco, Texas, USA.

Garcia, J., R. Kovner, and K. F. Green. 1966. Cue properties vs. palatability of flavors in avoidance learning. Psychonomic Science 20:313–314.

Gill, F. B. 1990. Ornithology. W. H. Freeman, New York, New York, USA.

Jacobs, G. H. 2009. Evolution of colour vision in mammals. Philosophical Transactions of the Royal Society B 364:2957–2967.

Kare, M. R., and J. G. Brand. 1986. Interaction of the chemical senses with nutrition. Academic Press, New York, New York, USA.

MacArthur, R. H., and E. R. Pianka. 1966. On the optimal use of a patchy environment. American Naturalist 100:603–609.

Mason, J. R. 1989. Avoidance of methiocarb-poisoned apples by red-winged blackbirds. Journal of Wildlife Management 53:836–840.

Mason, J. R., and L. Clark. 1997. Avian repellents: options, modes of action, and economic considerations. Pages 371–391 in J. R. Mason, editor. Repellents in wildlife management. National Wildlife Research Center, Fort Collins, Colorado, USA.

Mason, J. R., and L. Clark. 2000. The chemical senses in birds. Pages 39–56 in G. A. Whitow, editor. Sturkie's avian physiology. Fifth edition. Academic Press, New York, New York, USA.

Mason, J. R., and R. F. Reidinger. 1983. Generalization of and effects of pre-exposure on color-avoidance learning by red-winged blackbirds (Agelaius phoencieus). Auk 100:461–468.

Mason, J. R., J. A. Shivik, and M. W. Fall. 2001. Chemical repellents and other aversive strategies in predation management. Endangered Species Update 18:175–181.

Milgram, N. W., L. Krames, and T. M. Alloway. 1977. Food aversion learning. Plenum Press, New York, New York, USA.

Nachtman, T. J., J. H. Hull, and L. Clark. 2000. Water fog for repelling birds. U.S. Patent 6,024,971. Filed 7 April 1977. Issued 15 February 2000.

Nolte, D. L., J. R. Mason, G. Epple, E. Aronov, and D. L. Campbell. 1994. Why are predator urines aversive to prey? Journal of Chemical Ecology 20:1505–1516.

Pavlov, I. P. 1906. Conditioned reflex. Oxford University Press, Oxford, United Kingdom.

Pelchat, M. L., H. J. Grill, P. Rozin, and J. Jacobs. 1983. Quality of acquired responses to tastes by Rattus norvegicus depends on type of associated discomfort. Journal of Comparative Psychology 97:140–153.

Place, A. R., and E. W. Stiles. 1992. Living off the wax of the land: bayberries and warblers. Auk 109:334–345.

Provenza, F. D. 1995. Origins of food preference in herbivores. Pages 81–90 in J. R. Mason, editor. Repellents in wildlife management. National Wildlife Research Center, Fort Collins, Colorado, USA.

Reidinger, R. F. 1997. Recent studies on flavor aversion learning in wildlife damage management. Pages 101–120 in J. R. Mason, editor. Repellents in wildlife management. National Wildlife Research Center, Fort Collins, Colorado, USA.

Revusky, S. 1977. Learning as a general process with an emphasis on data from feeding experiments. Pages 1–71 in N. W. Milgram, L. Krames, and T. M. Alloway, editors. Food aversion learning. Plenum Press, New York, New York, USA.

Roper, T. J. 1999. Olfaction in birds. Pages 247–332 in P. J. B. Slater, J. S. Rosenblatt, C. T. Snowden, and T. J. Roper, editors. Advances in the study of behavior. Volume 28. Academic Press, New York, New York, USA.

Sayre, R., and L. Clark. 2001. Effect of primary and secondary repellents on European starlings: an initial assessment. Journal of Wildlife Management 65:461–469.

Shivik, J. A. 2004. Non-lethal alternatives for predation management. Sheep and Goat Research Journal 19:64–71.

Stager, K. E. 1964. The role of olfaction in food location by the turkey vulture (*Cathartes aura*). Los Angeles County Contributions to Science 81:1–63.

Stephens, D. W., J. S. Brown, and R. C. Ydenberg. 2007. Foraging: behavior and ecology. University of Chicago Press, Chicago, Illinois, USA.

Swihart, R. K., J. J. Pignatello, and M. J. I. Mattina. 1991. Aversive responses of white-tailed deer, *Odocoileus virginianus*, to predator urines. Journal of Chemical Ecology 17:767–777.

Teague, D. D. 2002. Vulture roost dispersal—improving air safety at Eglin AFB. Flying Safety 58:22–25.

Verheyden, C., and P. Jouventin. 1994. Olfactory behavior of foraging Procellariiformes. Auk 111:285–291.

Walsh, S., and A. Milner. 2011. Evolution of avian brain and senses. John Wiley and Sons, New York, New York, USA.

Werner, S. J., and L. Clark. 2003. Understanding blackbird sensory systems and how repellent applications work. Pages 31–40 *in* G. A. Linz, editor. Management of North American blackbirds. The Wildlife Society Wildlife Damage Management Working Group, Bismarck, North Dakota, USA.

Werner, S. J., G. M. Linz, J. C. Carlson, S. E. Pettit, S. K. Tupper, and M. M. Santer. 2011. Anthraquinone-based bird repellent for sunflower crops. Applied Animal Behaviour Science 129:162–169.

Zeigler, H. P., and H. J. Bischof. 1993. Vision, brain, and behavior in birds. Massachusetts Institute of Technology, Boston, USA.

4

Thomas W. Seamans
James A. Martin
Jerrold L. Belant

Tactile and Auditory Repellents to Reduce Wildlife Hazards to Aircraft

Wildlife within the airport environment are hazards to human safety. Lethal removal of targeted individuals reduces the immediate threat, but other approaches should be integrated into control programs to make them more effective and to help meet legal and ethical considerations (Dolbeer et al. 1995). When negative media attention, special interest groups, or calls for restrictive legislation influence public opinion, the resulting public pressure can preclude effective wildlife management and lead to subsequent population control problems (Torres et al. 1996, Coolahan and Snider 1998, Conover 2001). Nonlethal management activities to reduce wildlife use of airports may include habitat modification, exclusion from roosting and nesting areas, and repelling animals from desired locations. When considering repellents alone, there are many that are untested, temporarily effective, or cost-prohibitive (Dolbeer et al. 1995). Effective nonlethal repellents must affect some aspect of physical receptors or psychological perception of the intended targeted animals. In birds and mammals the primary physical receptors are visual (see Chapter 2), auditory, and tactile (Dooling 1982, Fay 1988, Clark 1998a). As explained in Chapter 3, the sense of smell is also important for birds and mammals. In this chapter we focus on auditory and tactile repellents, particularly the physiological bases for tactile and auditory repellent efficacy. We also examine some behavioral aspects of species that influence the efficacy of repellents.

Animal Sensory Capabilities

One must account for the auditory capability of animals when evaluating acoustic frightening devices. Auditory capabilities are measured in part by sound frequency in Hertz (Hz) and sound pressure level (SPL), the logarithmic measure of the pressure of a sound in decibels (dB) relative to a standard reference pressure in air (dB SPL), typically 20 µPa. Despite physical differences, the ears of mammals and birds work remarkably similarly. One obvious difference between the two groups is that avian ears are not externalized, yet have feather patterns that can focus sound waves into the ear in much the same way as the external mammalian ear. The avian inner ear differs from the mammalian inner ear, with one interior bone instead of three (Gill 2007). Even though the avian ear is structurally simpler than the coiled cochlea of a mammal, with its straight or slightly coiled cochlea (inner ear), the acoustical efficiency of birds is similar to that of mammals (Gill 2007). In both mammals and birds, hair cells in the cochlea serve as auditory sensory receptors. However, some birds, unlike mammals, have the ability to regrow some damaged hair cells (Ryals et al. 1999, Stone and Rubel 2000).

In general, birds hear well within a limited frequency range, whereas human hearing spans a wider range. Humans can detect sounds at frequencies from about 0.03 to 18 kHz (Heffner and Heffner 1992), with an absolute sensitivity at 0 dB SPL (Durrant and

Lovrinic 1984). Birds react most to sounds from 1 to 3 kHz, with an absolute sensitivity from −10 to 10 dB SPL (Dooling 1978, 1982; Stebbins 1983; Dooling et al. 2000). However, the range of sounds detected among species varies markedly. Downy woodpeckers (*Picoides pubescens*) are most sensitive to sounds from 1.5 to 4.0 kHz (Delaney et al. 2011), whereas barn owls (*Tyto alba*) are most sensitive from 6.0 to 7.0 kHz and at sound pressure levels as low as −18 dB SPL (Fay 1988). Rock pigeons (*Columba livia*) can detect low frequencies (0.05 Hz; i.e., in the infrasound range < 20 Hz), but it is unknown how pigeons use this capability (Fay and Wilber 1989, Fay and Popper 2000). Also, birds, unlike some mammals, do not hear ultrasonic (≥ 20 kHz) sounds (Schwartzkopff 1973, Dooling 1982).

White-tailed deer (*Odocoileus virginianus*), one of the most hazardous mammals to aircraft (Biondi et al. 2011, DeVault et al. 2011), hear from 0.25 to 54 kHz up to ~60 dB SPL. When measured using auditory brainstem response, deer were most sensitive to sounds from 4 to 8 kHz at 42 dB (D'Angelo et al. 2007). However, when measured using a behavioral audiogram, deer were most sensitive at 8 kHz and −3 dB SPL (Heffner and Heffner 2010). When measured at an intensity of 60 dB SPL, domestic dogs (Canidae) hear sounds between 0.067 and 44 kHz and domestic cats (Felidae) between 0.055 and 79 kHz (Heffner and Heffner 1992, Heffner 1998).

Hearing a sound and reacting to a sound require two different processes, however. Heffner (1998) describes these processes as sensation and perception, where sensation is the ability to detect a sound and perception is the ability to respond to the sound. This ability to respond is dependent not only on the physics of stimulus transmission, but also the ecological saliency of the stimulus (see Guilford and Dawkins 1991, Phelps 2007). We would expect an animal's perception to change as it habituates to a sound that is not negatively reinforced (see, however, Biedenweg et al. 2011). Vesper sparrows (*Pooecetes gramineus*) sometimes maintain breeding territories at airports despite noises associated with jet engines or passing vehicles. Yet vesper sparrows occupying territories in open fields that are not subject to constant airport noise, but are near roads, often cease singing and hide

when a car door is closed or a vehicle drives by (Seamans, personal observation; see also Summers et al. 2011).

In addition to auditory stimuli, animals perceive their environment through touch, primarily through contact with the skin. The skin of birds is relatively thinner than that of mammals, but as in mammals the skin serves multiple purposes. Skin provides a protective envelope for the body, some thermal insulation, and is a large sensory organ especially sensitive to temperature, pressure, and vibration (Stettenheim 1972, Schwartzkopff 1973). Although not apparent, the skin on a bird's foot is thick except at the hinges between the scales, where it is sensitive to tactile stimuli (Stettenheim 1972, Clark 1997). The trigeminal nerves in the avian bill are also sensitive to oral stimuli (Schwartzkopff 1973, Clark 1998*a*), which has been the basis for development of primary foraging repellents (e.g., methyl anthranilate–based products including Bird Shield and Bird Stop [Mason et al. 1989, Belant et al. 1996*b*, Clark 1998*b*]; Chapter 3). White-tailed deer have demonstrated sensitivity to electrical stimuli of 5.9 kV through their feet, a finding used in the development of electric mats as barriers against deer (Seamans and Helon 2008). Raccoons (*Procyon lotor*) have sensitive forepaws (Tremere et al. 2001) with good motor capability (Kaufmann 1982); therefore we assume they are reactive to tactile stimuli through their feet.

Premise for Efficacy

All vertebrates react to painful or noxious stimuli (Bateson 1991). Nonlethal techniques that cause direct pain or discomfort generally prompt animals to move away from the stimulus. However, both intra- and interspecific responses can vary depending on the situation, individuals involved, and type of stimulus (e.g., Hoffman and Fleshler 1965, Belant et al. 1997, Clark 1998*b*, Seamans and Blackwell 2011).

Most nonlethal management techniques are designed to evoke a response to a perceived predatory threat, which provides a strong motivation for animals to flee (Lima 1988, Keys and Dugatkin 1990, Lima and Dill 1990, Frid and Dill 2002). Flight from predators may be innate (Tinbergen 1948), learned, or enhanced via learning (Curio 1975, Kruuk 1976, Ydenberg and Dill 1986, Guilford 1990; see also Clark 1998*b*, Griffin 2004). A vast literature exists showing that prey response to a predator (i.e., antipredator behavior) varies due to numerous factors, including time of year in relation to breeding, frequency of predation risk, distance to escape cover, approach of the predator, type of habitat, and behavior of conspecifics (e.g., see Lima 1994, Cresswell et al. 2000, Elchuk and Wiebe 2002, Caro 2005, Devereux et al. 2006). In addition, humans represent a threat that generally elicits antipredator behaviors from most animals (e.g., see Bélanger and Bédard 1990, Evans and Day 2001, Frid and Dill 2002, Fernández-Juricic et al. 2003, Marzluff et al. 2010). Against birds, the efficacy of scare devices likely depends also on how targeted animals perceive stimuli relative to energy constraints and risk factors that affect foraging site selection (Suhonen 1993, Krams 1996, Elchuk and Wiebe 2002, Fernández-Juricic and Tran 2007).

The fate of animals frightened from a targeted area is often unknown but highly variable. Therefore the return rate of animals harassed or repelled from target areas offers a metric for method efficacy. Resident Canada geese (*Branta canadensis*) harassed from a park or residential area generally travel < 2 km (1.2 miles) and eventually return to the original site (Holevinski et al. 2007, Preusser et al. 2008). Aversive stimuli used against black bears (*Ursus americanus*), including nonlethal methods that caused pain, proved to be fairly ineffective in preventing bear returns to urban areas (Beckman et al. 2004). Even incidental disturbances of nontarget animals have demonstrated that other factors (i.e., nest defense) can override fear produced via novel stimuli. For instance, red-cockaded woodpeckers (*Picoides borealis*) returned to nests on average 4.4–6.3 min following the firing of 0.50-caliber blank rounds from a machine gun within 152 m (499 feet) of nests, but they did not leave nests when experiencing sound-exposure levels < 65 dB SPL at distances > 152 m (Delaney et al. 2011). Factors such as breeding season, availability of natural and anthropogenic food resources, and predation can clearly interact to diminish or enhance repellent effectiveness. An understanding of the context of application is critical in determining the types and necessary integration of repellent methods.

Auditory Repellents

Biosonic Stimuli

Auditory repellents are marketed as either ultrasonic, sonic, or biosonic calls. Human-made sounds are thought to frighten birds and therefore rely on the perception of danger (e.g., risk-disturbance hypothesis for nonlethal threats; Frid and Dill 2002). Loud (i.e., > 90 dB SPL) sounds may also cause physical distress. The underlying assumptions of biosonic recordings of bird alarm or distress calls are that (1) birds perceive such calls as natural warnings that danger is present and will subsequently flee (Lima and Dill 1990, Hurd 1996, Goodale and Kotagama 2008) and (2) birds are not as likely or will take longer to habituate to alarm and distress calls than other sounds (e.g., human-made sounds) because the calls are related to evolutionary signals of danger (Thompson et al. 1968, Johnson et al. 1985, Bomford and O'Brien 1990). Cliff swallow (*Petrochelidon pyrrhonota*) nesting activity was reduced 50% when alarm and distress calls were played in the nesting area (Conklin et al. 2009). Carrion crows (*Corvus corone*) responded to distress calls more than to effigies (Naef-Daenzer 1983). Likewise, Spanier (1980) found that about 88% of black-crowned night herons (*Nycticorax nycticorax*) left aquaculture facilities at the broadcasting of distress calls, and that no habituation was noted after six months. Researchers conjectured that herons that did not respond to distress calls were nonresident herons that had not established associations with conspecifics and therefore were not inclined to respond to the calls.

However, response to alarm calls may be species specific. Goodale and Kotagama (2008) found variation to response based on species ecology. Cook et al. (2008) found that unless a lethal element was added to distress calls, gulls (Laridae) habituated to the calls, whereas Coates et al. (2010) saw no response from wild turkeys (*Meleagris gallopavo*) to alarm calls. European starlings (*Sturnus vulgaris*) stopped responding to distress calls after about seven days when there was no negative reinforcement (Summers 1985). Additionally, call complexity may influence inter- and intraspecific responses (Soard and Ritchison 2009, Courter and Ritchison 2010, Fallow and Magrath 2010). Alarm and distress calls, though useful in bird control, are likely

Fig. 4.1. Biologist firing a pyrotechnic device from a specially designed pistol. Pyrotechnics are widely recognized as effective wildlife control tools when used as part of an integrated control program. Photo credit: Thomas W. Seamans

limited by context as well as by species behavior and ecology.

As discussed above, birds cannot hear ultrasonic sounds. Despite this fact, many ultrasonic devices are marketed as pest control devices. Bomford and O'Brien (1990) reviewed multiple studies that indicate ultrasonic stimuli are not aversive to birds, rodents, or insects. Although deer can hear in the ultrasonic range (D'Angelo et al. 2007, Heffner and Heffner 2010), in field trials they failed to react to ultrasonic devices, possibly because they were not loud enough for the deer to hear (Curtis et al. 1997, Belant et al. 1998, Valitzski et al. 2009).

Pyrotechnics

Pyrotechnics are auditory and visual devices that rely primarily on an explosion or other loud noise to scare birds (Mott 1980). The effect of a particular device might be sound alone or the combination of a particular sound with the light and smoke from the percussive component. Such devices include rifles and shotguns that fire live ammunition or blanks, or 12-gauge shotguns and flare pistols that fire exploding or noise-making projectiles (e.g., shell crackers, bird bombs, bird whistles, whistle bombs, or racket bombs; Fig. 4.1). The use of pyrotechnics to scare birds is widely recognized as an effective bird management tool (Booth 1994). The Humane Society of the United States (Hadidian et al. 1997) recognizes pyrotechnics as effective and humane scaring devices. Cleary and Dolbeer (2005) list the use of pyrotechnics as an effective means of reducing bird hazards at airports. However, some authors note that birds habituate to pyrotechnics and other scare devices (Blokpoel 1976, Inglis 1980, Slater 1980, Summers 1985). Limited lethal control has been suggested as a means to prolong the efficacy of pyrotechnic devices or to make the devices effective again after habituation occurs (Hochbaum et al. 1954, Slater 1980, Summers 1985, Smith et al. 1999), but limited empirical data have been provided to support this supposition. An exception is work by Baxter and Allan (2008), which showed that shooting some free-flying gulls at one feeding site enhanced the effectiveness of pyrotechnics, but that corvids did not respond similarly. Additionally, Cook et al. (2008) demonstrated that techniques including a lethal component were more effective at deterring birds than techniques with no lethal component. Killing one or more birds may provide the visual or auditory cue that stimulates a response by conspecifics (Guilford and Dawkins 1991). The presence of a dead bird alone can elicit a risk-avoidance response, but the perception of lethal attack might be a critical element for improving efficacy of effigies for some species (Avery et al. 2002, Seamans 2004, Seamans and Bernhardt 2004, Seamans et al. 2007b).

Exploders

Gas-operated exploders (e.g., gas cannons or propane cannons) have been commonly used since the late 1940s to repel pest birds from agricultural fields and airports (Gilsdorf et al. 2002). An exploder produces an extremely loud, intermittent explosion that exceeds the blast of a 12-gauge shotgun, which it is intended to simulate (Fig. 4.2). The assumption is that birds will associate the blast with gunfire and flee the area. Conover (1984) found exploders to be effective at reducing bird damage to corn, yet Washburn et al. (2006) found gulls at an airport did not respond to exploders even when lethal control with shotguns was conducted at the same site. Belant et al. (1996a) found that short-term responses of white-tailed deer to motion-activated exploders varied seasonally, but that regularly activated exploders were ineffective. As with other methods,

Fig. 4.2. Propane exploders have long been used to repel birds by simulating the sound of a shotgun blast, which is thought to represent a threat to birds. Photo credit: Thomas W. Seamans

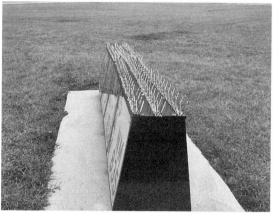

Fig. 4.3. Perching deterrents often include sharpened spikes affixed to areas attractive to birds; shown here is one such deterrent atop an airport sign. Photo credit: Todd Stewart

species-specific responses may vary depending on time of year, reproductive status, and environment in which the control tool is being used.

Tactile Repellents

Tactile repellents can be spikes of various designs, electric shock, tacky or sticky substances, moving or static wires, or chemical compounds designed to affect pain or discomfort (see also Chapter 3). In principle, all of these devices work by creating a painful or uncomfortable stimulus for the birds. Although many are listed for use against multiple bird species in numerous situations (Hygnstrom et al. 1994), few have been thoroughly tested.

Barriers

Spikes and wires in various arrangements are used in numerous situations, most often as barriers to deter birds from perching or loafing sites. Avery and Genchi (2004) evaluated the effectiveness of six different antiperching devices consisting of various arrangements of spikes, monofilament web, or a cone for deterring birds from perching. No single device was effective for all five species tested, as birds were able to find perching space that avoided contact with the spikes of some of the designs. Categorically, larger birds such as owls and vultures require different devices than do smaller species

(e.g., brown-headed cowbirds [*Molothrus ater*] and fish crows [*C. ossifragus*]). Seamans et al. (2007*a*) tested an antiperching wire and a spike-style device in an aviary setting, both of which were effective against European starlings, red-winged blackbirds (*Agelaius phoeniceus*), common grackles (*Quiscalus quiscula*), mourning doves (*Zenaida macroura*), and rock pigeons (Fig. 4.3).

Conklin et al. (2009) tested surface modifications in an effort to deter cliff swallows from nesting on highway structures. Polyethylene sheeting reduced nesting activity, although swallows were still able to build nests. However, silicon-based paint did not deter cliff swallows from nesting (Delwiche et al. 2010). Blocking ledges with sheeting or other materials placed at an angle of 45° or more also excluded birds from nesting and loafing areas (Williams and Corrigan 1994).

Chemical Applications

Reidinger and Libay (1979) reported that applying glue on perches near rice fields deterred birds for five to eight days. Belant (1993) found that roofs with tar surfaces reduced herring gull (*Larus argentatus*) nesting activity. Clark (1997) reported that starlings avoided perching on structures that had been treated with one of several dermal contact repellents that irritated the dermis on their feet, demonstrating agitation in response to 5% oil extracts of cumin, rosemary, and thyme. Furthermore, starlings avoided perches treated with

R-limonene, S-limonene, or β-pinene (Clark 1997). Products such as Hot Foot and Tanglefoot (polybutene-based repellents), although not based on the above extracts, are marketed as tactile repellents (Clark 1998b).

Electric Shock

The use of electric shock to keep wildlife from entering or using specific areas has seen limited field testing. However, the premise that an electrical stimulus is uncomfortable and that animals will avoid protected areas has been tested in numerous behavioral experiments with a wide variety of animals. Electric barriers of various designs have been used against white-tailed deer (Chapter 5). Smith et al. (1999) suggested using an electric fence to reduce Canada goose entrance into protected areas. Seamans and Blackwell (2011) found that an electrified perch repelled brown-headed cowbirds and rock pigeons, but that each individual had to experience the shock, as there was no apparent communication between flock members that deterred other individuals from the treated perch. Breck et al. (2006) developed an electrified repellent device that is activated by depressing a metal plate that completes a circuit. Although originally developed to deter black bears from concentrated food sources, it may deter other wildlife from spatially confined areas requiring protection. As long as an animal receives the uncomfortable sensation of an electric shock, it is likely that such a device will be effective (see Chapter 5 for further discussion). Unless a salient cue is provided with the stimulus, however, habituation to the shock could occur, as the electric impulse is not observable (Seamans and Blackwell 2011).

Compressed Air

Blasting air directly onto birds or through hoses that move rapidly and erratically also have been used to harass birds from roosting or loafing areas. When air is directed at high velocity and pressure, birds can be forced to move to alternative sites (White and Jinings 2006). In addition to the force of the air displacing some birds, the noise associated with airflow may displace birds that are not directly impacted, particularly as other flock members flee and alarm calls are sounded. Unlike instances when alarm calls alone are used and birds habituate to them (Summers 1985, Cook et al. 2008),

the physical displacement of birds should reinforce the alarm calls and reduce potential for habituation.

Summary

The principles behind auditory and tactile repellents are well founded in the biology of target species and behavioral ecology. The basis for the development of these repellents is the assumption that animals will flee and avoid treated areas in response to fear-provoking stimuli (e.g., alarm calls or explosive devices), physical barriers, or methods producing discomfort or pain. However, key components of these and other repellents include context application, association of the treatment with a negative outcome, and integration with other methods.

An understanding of context relative to method efficacy is especially important to controlling wildlife in airport environments, where a variety of noises (e.g., high-decibel engines) and visual stimuli (e.g., large moving objects, flashing lights), generally thought to be repellent to wildlife, are present and tolerated by birds and mammals. Research into the ultimate fate of animals after being targeted by a repellent would allow us to discern whether we are solving a problem, perhaps through dilution, or shifting the problem to a new site. Additionally, population studies to determine ultimate effects on survival of local populations following repeated control activities may be insightful to managers dealing with groups with opposing opinions about control activities.

LITERATURE CITED

Avery, M. L., and A. C. Genchi. 2004. Avian perching deterrents on ultrasonic sensors at airport wind-shear alert systems. Wildlife Society Bulletin 32:718–725.

Avery, M. L., J. S. Humphrey, E. A. Tillman, K. O. Phares, and J. E. Hatcher. 2002. Dispersing vulture roosts on communication towers. Journal of Raptor Research 36:45–50.

Bateson, P. 1991. Assessment of pain in animals. Animal Behaviour 42:827–839.

Baxter, A. T., and J. R. Allan. 2008. Use of lethal control to reduce habituation to blank rounds by scavenging birds. Journal of Wildlife Management 72:1653–1657.

Beckman, J. P., C. W. Lackey, and J. Berger. 2004. Evaluation of deterrent techniques and dogs to alter behavior of "nuisance" black bears. Wildlife Society Bulletin 32:1141–1146.

Bélanger, L., and J. Bédard. 1990. Energetic cost of man-induced disturbance to staging snow geese. Journal of Wildlife Management 54:36–41.

Belant, J. L. 1993. Nest-site selection and reproductive biology of roof- and island-nesting herring gulls. Transactions of the North American Wildlife and Natural Resources Conference 58:78–86.

Belant, J. L., S. K. Ickes, L. A. Tyson, and T. W. Seamans. 1997. Comparison of four particulate substances as wildlife feeding repellents. Crop Protection 16:439–447.

Belant, J. L., T. W. Seamans, and C. P. Dwyer. 1996a. Evaluation of propane exploders as white-tailed deer deterrents. Crop Protection 15:575–578.

Belant, J. L., T. W. Seamans, and L. A. Tyson. 1998. Evaluation of electronic frightening devices as white-tailed deer deterrents. Proceedings of the Vertebrate Pest Conference 18:107–110.

Belant, J. L., T. W. Seamans, L. A. Tyson, and S. K. Ickes. 1996b. Repellency of methyl anthranilate to pre-exposed and naïve Canada geese. Journal of Wildlife Management 60:923–928.

Biedenweg, T. A., M. H. Parsons, P. A. Flemming, and D. T. Blumstein. 2011. Sounds scary? Lack of habituation following the presentation of novel sounds. PLoS ONE 6:1–8, doi:10.1371/journal.pone.0014549.

Biondi, K. M., J. L. Belant, J. A. Martin, T. L. DeVault, and G. Wang. 2011. White-tailed deer incidents with U.S. civil aircraft. Wildlife Society Bulletin 35:303–309.

Blokpoel, H. 1976. Bird hazards to aircraft. Books Canada, Buffalo, New York, USA.

Bomford, M., and P. H. O'Brien. 1990. Sonic deterrents in animal damage control: a review of device tests and effectiveness. Wildlife Society Bulletin 18:411–422.

Booth, T. W. 1994. Bird dispersal techniques. Pages E19–E24 in S. E. Hygnstrom, R. M. Timm, and G. E. Larson, editors. Prevention and control of wildlife damage. University of Nebraska–Lincoln, Lincoln, USA.

Breck, S. W., N. Lance, and P. Callahan. 2006. A shocking device for protection of concentrated food sources from black bears. Wildlife Society Bulletin 34:23–26.

Caro, T. 2005. Antipredator defenses in birds and mammals. Chicago University Press, Chicago, Illinois, USA.

Clark, L. 1997. Dermal contact repellents for starlings: foot exposure to natural plant products. Journal of Wildlife Management 61:1352–1358.

Clark, L. 1998a. Physiological, ecological, and evolutionary bases for the avoidance of chemical irritants by birds. Pages 1–37 in V. Nolan and E. Ketterson, editors. Current ornithology. Volume 14. Plenum, New York, New York, USA.

Clark, L. 1998b. Review of bird repellents. Proceedings of the Vertebrate Pest Conference 18:330–337.

Cleary, E. C., and R. A. Dolbeer. 2005. Wildlife hazard management at airports: a manual for airport personnel. Second edition. Federal Aviation Administration, Office of Airport Safety and Standards, Washington, D.C., USA.

Coates, R., M. J. Delwiche, W. P. Gorenzel, and T. P. Salmon. 2010. Evaluation of damage by vertebrate pests in California vineyards and control of wild turkeys by bioacoustics. Human–Wildlife Interactions 4:130–144.

Conklin, J. S., M. J. Delwiche, W. P. Gorenzel, and R. W. Coates. 2009. Deterring cliff swallow nesting on highway structures using bioacoustics and surface modifications. Human–Wildlife Conflicts 3:93–102.

Conover, M. R. 1984. Comparative effectiveness of avitrol, exploders, and hawk-kites in reducing blackbird damage to corn. Journal of Wildlife Management 48:109–116.

Conover, M. R. 2001. Effect of hunting and trapping on wildlife damage. Wildlife Society Bulletin 29:521–532.

Cook, A., S. Rushton, J. Allan, and A. Baxter. 2008. An evaluation of techniques to control problem bird species on landfill sites. Environmental Management 41:834–843.

Coolahan, C. C., and S. Snider. 1998. Amendment 14—Colorado's anti-trapping initiative, a history and perspective on impacts. Proceedings of the Vertebrate Pest Conference 18:131–137.

Courter, J. R., and G. Ritchison. 2010. Alarm calls of tufted titmice convey information about predator size and threat. Behavioral Ecology 21:936–942.

Cresswell, W., G. M. Hilton, and R. D. Ruxton. 2000. Evidence for a rule governing the avoidance of superfluous escape flights. Proceedings of the Royal Society B 267:733–737.

Curio, E. 1975. The functional organization of anti-predator behaviour in the pied flycatcher: a study of avian visual perception. Animal Behaviour 23:1–115.

Curtis, P. D., C. Fitzgerald, and M. E. Richmond. 1997. Evaluation of the Yard Gard ultrasonic yard protector for repelling white-tailed deer. Proceedings of the Eastern Wildlife Damage Control Conference 7:172–176.

D'Angelo, G. J., A. R. De Chicchis, D. A. Osborn, G. R. Gallagher, R. J. Warren, and K. V. Miller. 2007. Hearing range of white-tailed deer as determined by auditory brainstem response. Journal of Wildlife Management 74:1238–1242.

Delaney, D. K., L. L. Pater, L. D. Carlile, E. W. Spadgenske, T. A. Beaty, and R. H. Melton. 2011. Response of red-cockaded woodpeckers to military training operations. Wildlife Monographs 177:1–38.

Delwiche, M. J., R. W. Coates, W. P. Gorenzel, and T. P. Salmon. 2010. Improved methods for deterring cliff swallow nesting on highway structures. Human–Wildlife Interactions 4:293–303.

DeVault, T. L., J. L. Belant, B. F. Blackwell, and T. W. Seamans. 2011. Interspecific variation in wildlife hazards to aircraft: implications for airport wildlife management. Wildlife Society Bulletin 35:394–402.

Devereux, C. L., M. J. Whittingham, E. Fernández-Juricic, J. A. Vickery, and J. R. Krebs. 2006. Predator detection and avoidance by starlings under differing scenarios of predation risk. Behavioral Ecology 17:303–309.

Dolbeer, R. A., N. R. Holler, and D. W. Hawthorne. 1995. Identification and control of wildlife damage. Pages 474–506 in T. A. Bookhout, editor. Research and management techniques for wildlife and habitats. The Wildlife Society, Bethesda, Maryland, USA.

Dooling, R. J. 1978. Behavior and psychophysics of hearing in birds. Journal of the Acoustical Society of America 64:S1–S4.

Dooling, R. J. 1982. Auditory perception in birds. Pages 95–130 in D. E. Kroodsma and E. H. Miller, editors. Acoustic communication in birds. Volume 1. Academic Press, New York, New York, USA.

Dooling, R. J., B. Lohr, and M. L. Dent. 2000. Hearing in birds and reptiles. Springer Handbook of Auditory Research 13:308–359.

Durrant, J. D., and J. H. Lovrinic. 1984. Basis of hearing science. Williams and Wilkins, Baltimore, Maryland, USA.

Elchuk, C. L., and K. L. Wiebe. 2002. Food and predation risk as factors related to foraging locations of northern flickers. Wilson Bulletin 114:349–357.

Evans, D. M., and K. R. Day. 2001. Does shooting disturbance affect diving ducks wintering on large shallow lakes? A case study on Lough Neagh, Northern Ireland. Biological Conservation 98:315–323.

Fallow, P. M., and R. D. Magrath. 2010. Eavesdropping on other species: mutual interspecific understanding of urgency information in avian alarm calls. Animal Behaviour 79:411–417.

Fay, R. R. 1988. Hearing in vertebrates: a psychophysics databook. Hill-Fay Associates, Winnetka, Illinois, USA.

Fay, R. R., and A. N. Popper. 2000. Evolution of hearing in vertebrates: the inner ears and processing. Hearing Research 149:1–10.

Fay, R. R., and L. A. Wilber. 1989. Hearing in vertebrates: a psychophysics databook. Journal of the Acoustical Society of America 86:2044.

Fernández-Juricic, E., A. Sallent, R. Sanz, and I. Rodríguez-Prieto. 2003. Testing the risk-disturbance hypothesis in a fragmented landscape: nonlinear response of house sparrows to humans. Condor 105:316–326.

Fernández-Juricic, E., and E. Tran. 2007. Changes in vigilance and foraging behaviour with light intensity and their effects on food intake and predator detection in house finches. Animal Behaviour 74:1381–1390.

Frid, A., and L. M. Dill. 2002. Human-caused disturbance stimuli as a form of predation risk. Conservation Ecology 6:11. http//www.consecol.org/vol6/iss1/art11.

Gill, F. B. 2007. Ornithology. Third edition. W. H. Freeman, New York, New York, USA.

Gilsdorf, J. M., S. E. Hygnstrom, and K. C. VerCauteren. 2002. Use of frightening devices in wildlife damage management. Integrated Pest Management Reviews 7:29–45.

Goodale, E., and S. W. Kotagama. 2008. Response to conspecifics and heterospecific alarm calls in mixed-species bird flocks of a Sri Lankan rainforest. Behavioral Ecology 19:887–894.

Griffin, A. S. 2004. Social learning about predators: a review and prospectus. Learning and Behavior 32:131–140.

Guilford, T. 1990. The evolution of aposematism. Pages 23–61 in D. Evans and J. O. Schmidt, editors. Insect defenses:

adaptive mechanisms and strategies of prey and predators. State University of New York Press, Albany, USA.

Guilford, T., and M. S. Dawkins. 1991. Receiver psychology and the evolution of animal signals. Animal Behaviour 42:1–14.

Hadidian, J., G. R. Hodge, and J. W. Grandy. 1997. Wild neighbors: the humane approach to living with wildlife. Humane Society of the United States. Fulcrum Publishing, Golden, Colorado, USA.

Heffner, H., Jr., and H. E. Heffner. 2010. The behavioral audiogram of whitetail deer (Odocoileus virginianus). Journal of the Acoustical Society of America 127:EL111–EL114.

Heffner, H. E. 1998. Auditory awareness. Applied Animal Behaviour Science 57:259–268.

Heffner, H. E., and R. S. Heffner. 1992. Auditory perception. Pages 159–184 in C. Phillips and D. Piggins, editors. Farm animals and the environment. C.A.B. International, Wallingford, United Kingdom.

Hochbaum, H. A., S. T. Dillon, and J. L. Howard. 1954. An experiment in the control of waterfowl depredations. Transactions of the North American Wildlife Conference 19:176–185.

Hoffman, H. S., and M. Fleshler. 1965. Stimulus aspects of aversive controls: the effects of response contingent shock. Journal of the Experimental Analysis of Behavior 8:89–96.

Holevinski, R. A., P. D. Curtis, and R. A. Malecki. 2007. Hazing of Canada geese is unlikely to reduce nuisance populations in urban and suburban communities. Human–Wildlife Conflicts 1:257–264.

Hurd, C. R. 1996. Interspecific attraction to the mobbing calls of black-capped chickadees (Parus atricapillus). Behavioral Ecology and Sociobiology 38:287–292.

Hygnstrom, S. E., R. M. Timm, and G. E. Larson. 1994. Birds. Pages E1–E3 in S. E. Hygnstrom, R. M. Timm, and G. E. Larson, editors. Prevention and control of wildlife damage. University of Nebraska–Lincoln, Lincoln, USA.

Inglis, I. R. 1980. Visual bird scarers: an ethological approach. Pages 121–143 in E. N. Wright, I. R. Inglis, and C. J. Feare, editors. Bird problems in agriculture. British Crop Protection Council, London, United Kingdom.

Johnson, R. J., P. H. Cole, and W. W. Stroup. 1985. Starling response to three auditory stimuli. Journal of Wildlife Management 49:620–625.

Kaufmann, J. H. 1982. Raccoon and allies. Pages 567–585 in J. A. Chapman and G. A. Feldhamer, editors. Wild mammals of North America. Johns Hopkins University Press, Baltimore, Maryland, USA.

Keys, G. C., and L. A. Dugatkin. 1990. Flock size and position effects on vigilance, aggression, and prey capture in the European starling. Condor 92:151–159.

Krams, I. A. 1996. Predation risk and shifts of foraging sites in mixed willow and crested tit flocks. Journal of Avian Biology 27:153–156.

Kruuk, H. 1976. The biological function of gulls' attraction towards predators. Animal Behaviour 24:146–153.

Lima, S. L. 1988. Initiation and termination of daily feeding in dark-eyed juncos: influences of predation risk and energy reserves. Oikos 53:3–11.

Lima, S. L. 1994. Collective detection of predatory attack by birds in the absence of alarm signals. Journal of Avian Biology 25:319–326.

Lima, S. L., and L. M. Dill. 1990. Behavioral decisions made under the risk of predation: a review and prospectus. Canadian Journal of Zoology 68:619–640.

Marzluff, J. M., J. Walls, H. N. Cornell, J. C. Withey, and D. P. Craig. 2010. Lasting recognition of threatening people by wild American crows. Animal Behaviour 79:699–707.

Mason, J. R., M. A. Adams, and L. Clark. 1989. Anthranilate repellency to starlings: chemical correlates and sensory perception. Journal of Wildlife Management 53:55–65.

Mott, D. F. 1980. Dispersing blackbirds and starlings from objectionable roost sites. Proceedings of the Vertebrate Pest Conference 9:38–42.

Naef-Daenzer, L. 1983. Scaring of carrion crows (*Corvus corone corone*) by species-specific distress calls and suspended bodies of dead crows. Proceedings of the Bird Control Seminar 9:91–95.

Phelps, S. M. 2007. Sensory ecology and perceptual allocation: new prospects for neural networks. Philosophical Transactions of the Royal Society B 362:355–367.

Preusser, S. E., T. W. Seamans, A. L. Gosser, and R. B. Chipman. 2008. Evaluation of an integrated non-lethal Canada goose management program in New York (2004–2006). Proceedings of the Vertebrate Pest Conference 23:66–73.

Reidinger, R. F., Jr., and J. L. Libay. 1979. Perches coated with glue reduce bird damage in ricefield plots. Bird Control Seminars 8:201–206.

Ryals, B. M., R. J. Dooling, E. Westbrook, M. L. Dent, A. MacKenzie, and O. N. Larsen. 1999. Avian species differences in susceptibility to noise exposure. Hearing Research 131:71–88.

Schwartzkopff, J. 1973. Mechanoreception. Pages 417–477 *in* D. S. Farner and J. R. King, editors. Avian biology. Volume 3. Academic Press, New York, New York, USA.

Seamans, T. W. 2004. Response of roosting turkey vultures to a vulture effigy. Ohio Journal of Science 104:136–138.

Seamans, T. W., S. C. Barras, and G. E. Bernhardt. 2007a. Evaluation of two perch deterrents for starlings, blackbirds and pigeons. International Journal of Pest Management 53:45–51.

Seamans, T. W., and G. E. Bernhardt. 2004. Response of Canada geese to a dead goose effigy. Proceedings of the Vertebrate Pest Conference 21:104–106.

Seamans, T. W., and B. F. Blackwell. 2011. Electric shock strips as bird deterrents: does experience count? International Journal of Pest Management 57:357–362.

Seamans, T. W., and D. A. Helon. 2008. Evaluation of an electrified mat as a white-tailed deer (*Odocoileus virginianus*) barrier. International Journal of Pest Management 54:89–94.

Seamans, T. W., C. R. Hicks, and K. J. Preusser. 2007b. Dead bird effigies: a nightmare for gulls? Proceedings of the 9th joint annual meeting. Bird Strike Committee–USA/Canada, 11 September 2007, Kingston, Ontario, Canada.

Slater, P. J. B. 1980. Bird behaviour and scaring by sounds. Pages 105–114 *in* E. N. Wright, I. R. Inglis, and C. J. Feare, editors. Bird problems in agriculture. British Crop Protection Council, London, United Kingdom.

Smith, A. E., S. R. Craven, and P. D. Curtis. 1999. Managing Canada geese in urban environments. Jack Berryman Institute Publication 16. Cornell Cooperative Extension, Ithaca, New York, USA.

Soard, C. M., and G. Ritchison. 2009. 'Chick-a-dee' calls of Carolina chickadees convey information about degree of threat posed by avian predators. Animal Behaviour 78:1447–1453.

Spanier, E. 1980. The use of distress calls to repel night herons (*Nycticorax nycticorax*) from fish ponds. Journal of Applied Ecology 17:287–294.

Stebbins, W. C. 1983. The acoustic sense of animals. Harvard University Press, Cambridge, Massachusetts, USA.

Stettenheim, P. 1972. The integument of birds. Pages 1–63 *in* D. S. Farner and J. R. King, editors. Avian biology. Volume 2. Academic Press, New York, New York, USA.

Stone, J. S., and E. W. Rubel. 2000. Cellular studies of auditory hair cell regeneration in birds. Proceedings of the National Academy of Sciences 97:11,714–11,721.

Suhonen, J. 1993. Predation risk influences the use of foraging sites by tits. Ecology 74:1197–1203.

Summers, P. D., G. M. Cunnington, and L. Fahrig. 2011. Are the negative effects of roads on breeding birds caused by traffic noise? Journal of Applied Ecology 48:1527–1534.

Summers, R. W. 1985. The effect of scarers on the presence of starlings (*Sturnus vulgaris*) in cherry orchards. Crop Protection 4:520–528.

Thompson, R. D., C. V. Grant, E. W. Pearson, and G. W. Corner. 1968. Cardiac response of starlings to sound: effects of lighting and grouping. American Journal of Physiology 214:41–44.

Tinbergen, N. 1948. Social releasers and the experimental method required for their study. Wilson Bulletin 60:6–51.

Torres, S. G., T. M. Mansfield, J. E. Foley, T. Lupo, and A. Brinkhaus. 1996. Mountain lion and human activity in California: testing speculations. Wildlife Society Bulletin 24:451–460.

Tremere, L., T. P. Hicks, and D. D. Rasmusson. 2001. Role of inhibition in cortical reorganization of the adult raccoon revealed by microiontophoretic blockade of GABA$_A$ receptors. Journal of Neurophysiology 86:94–103.

Valitzski, S. A., G. J. D'Angelo, G. R. Gallagher, D. A. Osborn, K. V. Miller, and R. J. Warren. 2009. Deer responses to sounds from a vehicle-mounted sound-production system. Journal of Wildlife Management 73:1072–1076.

Washburn, B. E., R. B. Chipman, and L. C. Francoeur. 2006. Evaluation of bird response to propane exploders in an airport environment. Proceedings of the Vertebrate Pest Conference 22:212–215.

White, R., and N. Jinings. 2006. Airport canopies become starling roosts—two airport case studies. Proceedings of the 8th joint annual meeting. Bird Strike Committee–USA/Canada, 21–24 August 2006, St. Louis, Missouri, USA.

Williams, D. E., and R. M. Corrigan. 1994. Pigeons (rock doves). Pages E87–E96 in S. E. Hygnstrom, R. M. Timm, and G. E. Larson, editors. Prevention and control of wildlife damage. University of Nebraska–Lincoln, Lincoln, USA.

Ydenberg, R. C., and L. M. Dill. 1986. The economics of fleeing from predators. Advances in the Study of Avian Behaviour 16:229–249.

5 — Excluding Mammals from Airports

Kurt C. VerCauteren
Michael Lavelle
Thomas W. Seamans

To ensure aircraft safety, it is critical to exclude large mammal species such as deer (*Odocoileus* spp.), feral swine (*Sus scrofa*), and coyotes (*Canis latrans*) from airport environments, as well as to consider thoroughly and carefully all available management methods. Airports are often located on or adjacent to undeveloped land that provides habitat for various species large enough to pose a direct hazard to aircraft. Unoccupied expanses of forage near runways provide deer with sufficient incentive to leave cover and occupy airport lands. Associated risk and tragic collisions have ranked deer as the most hazardous wildlife group to aviation (Dolbeer et al. 2000, DeVault et al. 2011), necessitating the evaluation of appropriate means for excluding them and other medium to large mammals (Dolbeer et al. 2000). Exclusionary fences are the most effective, long-lasting, and straightforward tool for eliminating risks posed by deer and other large mammals at airports; however, these fences can be costly to purchase, erect, and maintain. Fences provide a visual sense of security for airport managers but also can accomplish a measurable and statistically significant level of protection to aircraft at airports (DeVault et al. 2008). A variety of evaluations and experiments have been conducted on fence options. Determining the most appropriate fence for a specific setting to accomplish a desired outcome can be challenging. When reviewing this body of literature, airport managers must consider the level of motivation among deer or other species in the experiment and relate it to their situation. In this chapter we review a variety of fence applications for excluding medium to large mammals and provide recommendations.

Federal Aviation Administration Recommendations

The Federal Aviation Administration (FAA) prepares and circulates advisories on recommended practices to airport operators and safety inspectors. Since 2000, the FAA has disseminated three particular advisories, called CertAlerts, related to fencing strategies for deer (see http://www.faa.gov/airports/airport_safety/certalerts/). The first (No. 01-01; Castellano 2001) established minimum fence standards for excluding deer from airports. Standards specified chain-link fence at least 2.4 m (8 feet) high with 0.6-m (2-foot) outriggers with an unspecified number of strands of barbed wire. Recommendations specify that the fence must also be buried a minimum of 0.6 m (2 feet) and monitored daily. In 2004, recommendations were revised to specify a 3.0-m (10-foot) chain-link fence topped with three strands of barbed wire and a 1.2-m (4-foot) skirt buried in the ground at a 45° angle on the outside of the fence (Castellano 2004).

Research results compiled by the National Wildlife Research Center, which is part of the U.S. Department of Agriculture, Wildlife Services program, prompted the release of CertAlert No. 02-09, stating that alternative electric-fence designs (1.2–1.8 m [4–6 feet] high, 5–9 strands) proved 99% effective in stopping deer

Fig. 5.1. Breaches in airport fencing can allow easy access to the air operations area. From DeVault et al. (2008)

and could be suitable in limited, though unspecified, situations at airports (Castellano 2002). In 2004, an additional CertAlert was released that included all of the above information but specified that gates in fence lines must provide no more than a 15.2-cm (6-inch) gap that could potentially allow access by deer (Castellano 2004). Minimum recommendations provided in the CertAlerts for chain-link fences are appropriate when land managers must virtually eliminate access by medium to large mammals, realizing there is always potential for a break in a fence to occur by uncontrollable causes.

Deer-Strike Statistics

From 1990 through 2009, the FAA received 964 reports of deer–aircraft collisions (i.e., deer strikes)—including white-tailed deer (*O. virginianus;* 879), mule deer (*O. hemionus;* 55), and generic "deer" of undetermined species (30)—with 84% of the strikes resulting in damage (Dolbeer et al. 2011). Reported cost of the strikes was $31.7 million (http://wildlife-mitigation .tc.faa.gov/wildlife/database.aspx). Coyotes are an additional wildlife hazard, resulting in 321 strikes, 22% of which having an adverse effect on aircraft and 9% causing damage (Dolbeer et al. 2011). As populations of deer and feral swine continue to increase (Côté et al. 2004, Ditchkoff and West 2007, respectively) the threat of strikes increases, mandating the exclusion of these mammals from airports.

Physical Abilities

When attempting to exclude or contain an animal, its size, intelligence, and physical ability must be considered (Fitzwater 1972). There are a variety of published studies that evaluate fence designs capable of excluding various-sized wildlife, including small rodents (e.g., Connolly et al. 2009, Honda et al. 2009). Mammals may get past a fence by going over, under, or through it (Fig. 5.1).

When we focus on jumping ability, for example, we find that literature and observations suggest deer are capable of jumping 2.3- to 2.4-m (7.5- to 8-foot) fences (Falk et al. 1978, Sauer 1984) and that fences < 3 m (10 feet) high might not be entirely deer proof (Curtis et al. 1994, Kaneene et al. 2002, VerCauteren et al. 2006a). Yet documented cases of deer penetrating such fences are scarce in published literature, so researchers sought to verify the true abilities of white-tailed deer by conducting a series of experiments in which they motivated deer to jump progressively higher fences until they would jump no higher (VerCauteren et al. 2010). Deer in their study would not jump a 2.4-m fence, and very few (< 10%) would jump 2.1 m (7 feet), suggesting that a 2.4-m fence will contain or exclude most white-tailed deer (VerCauteren et al. 2010). However, incidental observations of deer jumping 2.4-m fences (see Arnold and Verme 1963, Sauer 1984) indicate that a well-constructed and maintained fence of > 2.4 m in height is justified where 100% deterrence is required, such as at airports.

Deer are not only adept at jumping barriers but are more likely to maneuver through or under poorly constructed fences (Feldhamer et al. 1986). Black bears (*Ursus americanus*) are proficient climbers and have been documented climbing 1.8-m (6-foot) fences, presenting yet another challenging species to exclude (deCalesta and Cropsey 1978). Coyotes are capable of jumping 1.5-m (5-foot) fences from a standstill and can climb 1.8-m wire-mesh fences (Thompson 1978; Fig. 5.2). Burying fences or installing aprons of wire mesh on fences, as suggested in FAA CertAlerts (Castellano 2001, 2002, 2004), not only reduces potential for burrowing animals digging under a fence, but also minimizes risk of other larger mammalian species entering beneath a fence (Fig. 5.3).

Fig. 5.2. Coyote scaling a fence at a major western U.S. airport. Photo credit: Port of Portland

Openings in fences that appear small enough to impede deer may actually be large enough for motivated deer or other mammals to squeeze through. Adult white-tailed deer were able to pass through a 25.0-cm (10-inch) gap at the bottom of a fence (Falk et al. 1978, Palmer et al. 1985, Feldhamer et al. 1986). Caribou (*Rangifer tarandus*) will also pass through a fence rather than jump, even though they are capable of jumping 2.2 m (7.5 feet; Miller et al. 1972). Coyotes are capable of crawling through 15.2 × 10.2 cm (6 × 4 inch) openings and can walk through 30.5-cm (12-inch) mesh (Thompson 1978). Ward (1982) reported that a 15.0-cm (6-inch) gap under a fence was enough to allow passage by mule deer, and Feldhamer et al. (1986) documented deer in Pennsylvania, USA, passing through 19.0-cm (7.5-inch) openings. Ultimately, a fence must be of sufficient height, tight to the ground or preferably buried, and lack gaps >15.0 cm² (2.3 inches²) to ensure exclusion of deer.

Fig. 5.3. Some mammals, including coyotes, can penetrate fencing without a belowground apron by burrowing. From DeVault et al. (2008)

Motivational Factors

Overall efficacy of fences for impeding passage is usually related directly to the associated level of motivation. As such, the more motivated an animal is to penetrate a fence, the more substantial the fence needs to be (Goddard et al. 2001). Deer and other animals stressed by immediate life-or-death situations (e.g., being pursued by a predator) frequently exhibit atypical behaviors and, under certain circumstances, may penetrate a fence that would otherwise deter them (Bryant et al. 1993, Conover 2002, Lavelle et al. 2011). The motivation to vacate lands adjacent to an airport may be unpredictable, supporting the need for robust fence construction in such areas. Complete enclosure of airports is justified, though not all areas (i.e., adjacent to areas with minimal human activity) require the same level of security.

Motivational factors such as seasonal and daily movements, food, and predators (including humans) are important considerations in assessing the efficacy of a fence design. For example, deer collisions with aircraft peak in October and November (Biondi et al. 2011), as do collisions with automobiles, a direct correlation to increased movements associated with the breeding season (Bellis and Graves 1971, Hawkins et al. 1971). Most collisions occur during crepuscular periods when deer activity peaks or at night, when lowered visibility makes deer detection more difficult (Carbaugh et al. 1975, Biondi et al. 2011).

If food is abundant and competition minimal, deer will be less motivated to access resources on the other side of a barrier, suggesting that a less substantial fence design may be adequate and effective (DeNicola et al. 2000), depending on the need and consequences of a breach. For example, under minimal motivation, simple fencing such as a 25-cm single-strand electric fence can be effective in excluding deer (Steger 1988). Deer with slightly more motivation were excluded from a 4-ha melon planting with the use of a four-strand electric fence that was 97.0 cm (3.2 feet) tall, resulting in the producer's first harvestable crop in years (McAninch 1986). Complicating the issue further, individuals competing for food will try harder to penetrate a fence to access food on the other side.

Factors Contributing to Breaches

Habitat adjacent to a fence also influences the level of motivation to breach that fence. Feldhamer et al. (1986) examined the efficacy of two fence designs for excluding deer, including a 2.7-m (9-foot) woven-wire fence and a 2.2-m (7.2-foot) woven-wire fence topped with two additional strands of high-tensile wire along an interstate highway. When adjacent to forested areas, the 2.7-m fence was more effective than the 2.2-m fence, but on nonforested or level ground, efficacy between fences was similar. Deer rarely, if ever, attempted to jump the 2.7-m fence, choosing instead to go under wherever possible.

One difficultly encountered in fence installation is inflexible fence material that cannot follow ground contours, resulting in gaps between the fence and the ground. A single strand of barbed or high-tensile wire strung below a fence can be a simple solution to the problem of gaps (Bryant et al. 1993). Gaps can also be avoided by investing time and money to create a straight and level course for fence installation, improving overall efficacy and visibility of the fence to approaching animals and minimizing damage from falling trees and limbs (Smith 1983, Palmer et al. 1985).

Current recommendations for airport fence construction include the addition of a 1.2-m apron extending underground at a 45° angle on the outside of the fence (Castellano 2004). Ideally, this addition would be made as the fence is being constructed, though it could be added to an existing fence. The addition of an apron will all but eliminate potential for deer, coyotes, and most other medium to large mammals to enter an airport by passing under a fence (Fig. 5.3).

Economics of Fencing

Aircraft strikes with medium to large mammals are costly in terms of damage to equipment and potential for injuries or death to humans (Biondi et al. 2011, DeVault et al. 2011). As a result, it seems sensible to provide maximum protection for all airports. If the presence of medium to large mammals is not acceptable, airports should accept the cost and erect the most substantial fence available. We realize, however, that smaller, noncommercial airports may be financially limited and that erecting a less extensive fence than recommended by the FAA may be the only option. Consequently, smaller airports often have varying levels of perimeter fencing that reveal vulnerabilities to threats posed by deer and other mammals (DeVault et al. 2008). Although cost ultimately determines which means for exclusion is chosen, construction details are also important. DeVault et al. (2008) documented a situation in which deer followed a fence of suitable height and configuration but of insufficient length. Deer traveled to its end, where they gained access to an airfield and corn that was available on the other side of the runway.

Airports frequently cover large expanses of land, oftentimes requiring the installation of kilometers of fence. The relationships between size and shape of the area being fenced, and how they affect total costs, should also be considered (VerCauteren et al. 2006a). Larger areas are more cost-effective (lower cost per unit area) to fence than smaller areas, because as the perimeter length increases, the area enclosed increases to a greater degree (Brenneman 1983, McAninch 1986). Further, square areas are more cost-effective to enclose than elongated or oddly shaped areas of the same size.

When weighing the merits of installing a fence to control deer damage to crops, the cost relative to the fence's potential savings should be considered. Until recently, efforts to manage wildlife damage have rarely been evaluated economically (Caughley 1977, Dyer and Ward 1977, Caslick and Decker 1979, Dolbeer 1988, Blackwell et al. 2003). Researchers have placed more emphasis on determining statistical significance of experiments than on evaluating economic significance (Dillon 1977). Yet economic modeling of systems related to risks posed by wildlife is worthwhile and important in considering management strategies (VerCauteren et al. 2002). In situations where economic benefit can be quantified, economic models can facilitate selection of fence type to be used (VerCauteren et al. 2006b). Net present values can be used to determine which type of fence, if any, would be cost-effective. Net present values compare the value of a dollar today to its value in the future and is an efficient way to measure benefits and costs that accrue over the lifetime of a particular fence design. A model on fence selection related to deer damage provides users with tools to make informed decisions regarding fencing options (VerCauteren et al. 2006b). This best fence selection model provides economic analyses and predicts the economic outcomes relative to the area and perimeter of the protected area, value of the resource being protected, cost, life span, and efficacy of the fence. The model can increase user awareness regarding how parameters such as efficacy or life span fluctuate with varying level of financial investment, and it may provide insight for airport managers tasked with selecting the best fence for the situation.

Although all fences require regular maintenance to remain effective, inexpensive fences like the baited electric version require additional maintenance in application of attractants or repellents. A less expensive fence may require more maintenance and may not last as long as a fence that requires a higher initial investment (Byrne 1989). Current FAA recommendations specify the need for daily fence checks to eliminate the possibility of allowing access to airports, and appropriate labor estimates should be incorporated into predicted budgets for fencing applications.

Fence Options

Fences exclude or contain animals by providing a physical barrier, a psychological barrier (via behavioral conditioning), or a combination of both. Fences such as woven wire present a physical barrier that prevents animals from passing over, through, or under. Conversely, a two-strand electric polytape fence provides a minimal physical barrier but acts as a psychological barrier by delivering negative stimuli (shock) upon contact (McKillop and Sibly 1988, Curtis et al. 1994). Other fences, like electric 15-strand high-tensile wire, function by combining both effects. Traditionally, fences of wire-mesh construction were used for excluding or containing deer and other mammals. More recently, electric fences consisting of multiple strands of high-tensile steel wire or polyrope have gained popularity, as associated costs and labor are lower than traditional wire-mesh fences (VerCauteren et al. 2006a; Fig. 5.4). They require additional vigilance and maintenance, however.

Temporary Fences

Although many fences are erected as long-term installations—providing protection for >30 years with reg-

Fig. 5.4. Novel fencing designs include electrified polyrope and electrified mats that allow the passage of vehicles through gates but deter mammals. Photo credit: Kurt C. VerCauteren

ular maintenance—risk of damage is often seasonal, related to periodic factors such as migration, accessing preferred foods (Flyger and Thoerig 1962), and breeding season (Marsh et al. 1990). In situations when year-round protection may not be deemed necessary, a variety of temporary fence designs, such as polytape and polypropylene snow fence, may be sufficient. When protecting particular agricultural resources (e.g., ripening crops, orchards, etc.), the need for protection may be only temporary. Though surely limited, there may be airports where only seasonal protection is needed (e.g., migrating caribou herds), and temporary fences might fulfill that need.

Temporary fences may be less expensive, but they are also less durable and less effective than permanent fences and may be prone to damage and degradation. Temporary fences are typically lightweight (i.e., polypropylene, nylon) and often erected using posts that do not involve digging and can be installed with handheld post pounders. Steel T-posts or fiberglass posts are sufficient for most temporary fence installations.

Electric Fences

Although other types of fences physically keep wildlife out of airports, electric fences typically rely on behavioral conditioning by delivering a shock to animals attempting to breach them (Porter 1983, McKillop and Sibly 1988, Curtis et al. 1994, Leblond et al. 2007). At airports where deer densities and motivation to enter are minimal and smaller mammals are not a concern, electric fences may be entirely adequate, though they have limitations. Electric multistrand, high-tensile wire or electric polyrope fences are comparably priced at $4 to $13/m ($1 to $4/foot) installed (Seamans and VerCauteren 2006) but are typically less effective than wire-mesh fences because their deterrence relies solely upon delivery of negative stimuli (McKillop and Sibley 1988).

For electric fences, two general rules apply: first, erect them before animals are in the habit of entering the area (Wilson 1993, Craven and Hygnstrom 1994, Curtis et al. 1994) and second, keep the fence electrified. If a fence loses power, animals like deer will be quick to penetrate it (Ward 1982, Clevenger et al. 2001, Conover 2002, Poole et al. 2004). Additionally, failure-detection devices should be incorporated into electric fence systems to minimize potential for breaches and to allow for prompt repairs (Leblond et al. 2007). Other factors should be taken into account when considering the use of electric fences, including voltage requirements, charge configuration, fence configuration, seasonal fences, and attractants. For successful deer control, high-tensile wire and polytype materials should carry a minimum charge of 3,000 V (Matschke et al. 1984, Duffy et al. 1988, Curtis et al. 1994). Fence design should reflect the size of target species to ensure wire spacing is sufficient to deliver adequate charge to offending animals, such as strand spacing no greater than 15.2 cm for deer. Also, electric fences are most effective when target individuals approach calmly and slowly, receiving a significant shock that prompts retreat. Fences that allow wildlife to approach with the momentum to carry them through the barrier are not as effective (McKillop and Sibley 1988).

Various materials are available for constructing electric fences. The most durable and longest-lasting option is high-tensile strength, smooth steel wire and is commonly available in 12.5-gauge natural galvanized and green colorations. Such fences have been used to contain and exclude large mammals in New Zealand for nearly 40 years (Byrne 1989). Numerous field trials have shown that they have nearly eliminated passage by deer (Tierson 1969, Brenneman 1982, Palmer et al. 1985). Craven and Hygnstrom (1994) reported slanted and upright high-tensile fences to be suitable for protection of orchards, large vegetable gardens, and other fields under moderate to high deer pressure, whereas the offset electric may only be suitable for smaller fields (<1.6 ha) under moderate deer pressure. Average costs of materials to construct a high-tensile electric fence range from $2 to $5/m ($0.6 to $1.5/foot). Proper maintenance requires frequent inspection, seasonal tensioning of wire, and suppression of vegetation. Electric high-tensile fences may not offer the same security as wire-mesh fences of comparable height, but they can be less expensive. Fallen trees, for example, will occasionally compromise a fence, but the elasticity of high-tensile wires often keeps them from breaking, and they often spring back into place once trees are removed (Brenneman 1983). Although cost and characteristics may be appealing, when 100% exclusion is necessary, these fences should not be considered.

The integration of petroleum-derived woven materials (primarily polypropylene) and strands of conductive metal wires has revolutionized the fencing industry. Polyrope, polywire, polytape, and polynet fences are widely available and appropriate for a variety of applications. Polyrope, such as that developed by ElectroBraid Fence Ltd. (Yarmouth, Nova Scotia, Canada; see also Seamans and VerCauteren 2006), is now an acceptable option in some airport environments and can be installed closer to areas of aircraft movement than traditional wire-mesh fences (Castellano 2002). These polyfence options are particularly appealing over wire options because of their easier construction, teardown, and storage if only used seasonally, as well as their high visibility and potentially increased efficacy against approaching wildlife, which may minimize animal–fence collisions (Hygnstrom and Craven 1988). Additionally, electric fences of polyrope construction can significantly reduce movements by moose (*Alces alces*; Leblond et al. 2007) and feral swine (Reidy et al. 2010), though they are by no means impenetrable. Managers can minimize problems with vegetation shorting-out these fences by using low-impedance energizers or by running positive and negative charges on alternating strands.

Wire-Mesh Fences

In general, fences of wire-mesh construction are installed with the expectation of a long and effective life span (Isleib 1995). This is often exactly what is needed in an airport setting, and so woven-wire mesh fence designs are well suited. Areas requiring high security (i.e., airports and correctional facilities) necessitate substantial fence heights in excess of 2.4 m, which are available in various wire-mesh construction, including woven wire, chain link, and V-mesh, but these options are not created equal. Wire-mesh materials vary in weight, durability, expected life span, ease of construction, and cost. Woven-wire fence was favored by survey respondents in Michigan and Wisconsin and considered very effective for excluding deer from crops (Isleib 1995). Quality wire-mesh fence materials cost $10 to >$20/m ($3 to >$6/foot) and can last >30 years (Curtis et al. 1994).

Chain link is frequently the material of choice for airport installations. As such, recommendations for airports mainly emphasize use of chain-link fence of 2.4 m topped by additional fence materials or 3.05 m in height. Chain link is typically perceived as providing the highest-level security with minimally spaced mesh, enabling it to be effective in excluding all but smaller mammals.

Other wire-mesh fence designs similar to chain link include high-tensile woven wire, welded-wire mesh, and V-mesh. Each material has advantages and disadvantages, but for airports that need to exclude animals from the size of fox to moose, chain-link fencing is the most desirable fencing material. Woven-wire mesh is typically less expensive and easier to install than chain link, but its larger mesh spacing also makes it less effective for excluding young animals (Lavelle et al. 2011). Likewise, wire-mesh fence is commonly used to minimize wildlife–vehicle collisions along busy highways within migration corridors. A 2.4-m wire-mesh fence along highways can be effective in reducing wildlife collisions, especially when used in conjunction with alternate routes of passage that allowed for continued movement while minimizing motivation to breach a fence (Ward 1982, Lehnert and Bissonette 1997, Clevenger et al. 2001).

Gates

Traditional hinged gates constructed of materials at least as stout and tall as adjoining fence lines provide comparable levels of protection; however, in high-traffic areas they may not be practical. In low-traffic areas, gates may be considered a nuisance and are potentially left open, creating risk by allowing entry by wildlife (Seamans 2001). Open gates are often the cause for animals ending up where they should not be (Van Noord 2000). Alternatives to traditional gates are being developed and tested, both with scientific rigor and in ongoing management practices (Bashore and Bellis 1982, Seamans and Helon 2008, VerCauteren et al. 2009). Means to allow easy access by vehicles and machinery while effectively preventing passage of medium to large mammals are needed. VerCauteren et al. (2009) compared commercially available Bumpgates, novel deer guards (multiple conveyors placed over a pit 0.4 m [1.3 feet] deep), and unprotected plots, and demonstrated that alternatives to traditional gates exist; however, these alternatives may not be suitable for high-security applications where any entries are unacceptable. Reed et al. (2007) tested modified cattle guards that were 3.7 m (12 feet), 5.5 m (18 feet), and 7.3 m (24 feet) long for controlling movements of deer, with little success (16 of 18 deer monitored successfully crossed the guard). Peterson et al. (2003) also evaluated three designs of deer guards and found bridge grating to be 99% effective at excluding Key deer (*O. v. clavium*). Belant et al. (1998) developed a design with round tubing and successfully excluded >88% of deer, compared to pretreatment crossings. Seamans and Helon (2008) evaluated the use of experimental electric mats (Fig. 5.4) as an alternative to gates and found them to be 95% effective. At airports, bridge grates or electric mats in conjunction with hinged gates that are closed during times of low traffic volume may be excellent options.

Gates are not only necessary to eliminate passage by medium to large mammals into or out of an area, they may play an important role in allowing them to exit an area from which they were intended to be excluded. Managers should proactively prepare for unforeseen occurrences where animals inadvertently access airports. One way is to construct devices (i.e., one-way gates, earthen escape ramps) that allow animals that entered to exit on their own without human intervention. For example, one-way gates, constructed of a funnel-like assemblage of metal tines, were developed and evaluated for allowing deer to exit highway rights-of-way (D'Angelo et al. 2007, Reed et al. 2007).

Although they may be only occasionally effective, they are routinely used in large fence installations. In comparing one-way gates to earthen escape ramps, ramps were roughly ten times more effective in enabling deer to exit highway rights-of-way (Bissonette and Hammer 2000). Stull et al. (2011) found woven-wire fence topped with outriggers angled away from the protected area acted as one-way barriers allowing animals to exit easier than entering.

Summary

Of all available methods for alleviating potential risk of aircraft–mammal collisions at airports, exclusionary fencing is the most straightforward, effective, recommended, and most used. Even so, costs for supplies, construction, and maintenance can seem prohibitive. When considering the level of security needed to exclude deer and other mammals from airports, managers must ask, is anything short of 100% exclusion acceptable? When human lives are at stake, erecting one of the many effective varieties of exclusionary fencing is imperative. Selection of appropriate fence materials should involve consideration of multiple factors, including level of acceptable risk, maximum potential levels of motivation of deer and other mammals to breach, surrounding habitat types, seasonality of hazards, and costs (both in supplies and labor for the life span of the fence). Although erecting and maintaining an exclusionary fence may seem like the complete solution to medium and large mammal–related hazards at airports, management of these hazards should allow for additional strategies to be implemented as needed. Population management strategies (Chapter 7) may be necessary on adjacent lands to minimize pressure for animals to enter airport properties. Additionally, plans for use of frightening devices and lethal management tools should be established in the event of a fence breach. Any technique can fail, so mitigation measures must be immediately available to minimize potential for disaster.

LITERATURE CITED

Arnold, D. A., and L. J. Verme. 1963. Ten year's observation of an enclosed deer herd in northern Michigan. Transactions of the North American Wildlife Conference 28:422–430.

Bashore, T. L., and E. D. Bellis. 1982. Deer on Pennsylvania airfields: problems and means of control. Wildlife Society Bulletin 10:386–388.

Belant, J. L., T. W. Seamans, and C. P. Dwyer. 1998. Cattle guards reduce white-tailed deer crossings through fence openings. International Journal of Pest Management 44:247–249.

Bellis, E. D., and H. B. Graves. 1971. Deer mortality on a Pennsylvania interstate highway. Journal of Wildlife Management 35:232–237.

Biondi, K. M., J. L. Belant, J. A. Martin, T. L. DeVault, and G. Wang. 2011. White-tailed deer incidents with U.S. civil aircraft. Wildlife Society Bulletin 35:303–309.

Bissonette, J. A., and M. Hammer. 2000. Effectiveness or earthen return ramps in reducing big game highway mortality in Utah. UTCFWRU Report Series 2000(1):1–29.

Blackwell, B. F., E. Huszar, G. Linz, and R. A. Dolbeer. 2003. Lethal control of red-winged blackbirds to manage damage to sunflower: an economic evaluation. Journal of Wildlife Management 67:818–828.

Brenneman, R. 1982. Electric fencing to prevent deer browsing on hardwood clearcuts. Journal of Forestry 80:660–661.

Brenneman, R. 1983. Use of electric fencing to prevent deer browsing in Allegheny hardwood forests. Proceedings of the Eastern Wildlife Damage Control Conference 1:97–98.

Bryant, L. D., J. W. Thomas, and M. M. Rowland. 1993. Techniques to construct New Zealand elk-proof fence. General Technical Report PNW-GTR-313. U.S. Department of Agriculture Forest Service, Pacific Northwest Research Station, Portland, Oregon, USA.

Byrne, A. E. 1989. Experimental applications of high-tensile wire and other fencing to control big game damage in northwest Colorado. Proceedings of the Great Plains Wildlife Damage Control Workshop 9:109–115.

Carbaugh, B., J. P. Vaughan, E. D. Bellis, and H. B. Graves. 1975. Distribution and activity of white-tailed deer along an interstate highway. Journal of Wildlife Management 39:570–581.

Caslick, J. W., and D. J. Decker. 1979. Economic feasibility of a deer-proof fence for apple orchards. Wildlife Society Bulletin 7:173–175.

Castellano, B. 2001. Deer aircraft hazard. CertAlert No. 01-01. Federal Aviation Administration, Airports Safety and Operations Division, Washington, D.C., USA.

Castellano, B. 2002. Alternative deer fencing. CertAlert No. 02-09. Federal Aviation Administration, Airports Safety and Operations Division, Washington, D.C., USA.

Castellano, B. 2004. Deer hazard to aircraft and deer fencing. CertAlert No. 04-16. Federal Aviation Administration, Airports Safety and Operations Division, Washington, D.C., USA.

Caughley, G. 1977. Analysis of vertebrate populations. John Wiley and Sons, London, United Kingdom.

Clevenger, A. P., B. Chruszcz, and K. E. Gunson. 2001. Highway mitigation fencing reduces wildlife–vehicle collisions. Wildlife Society Bulletin 29:646–653.

Connolly, T. A., T. D. Day, and C. M. King. 2009. Estimating the potential for reinvasion by mammalian pests through pest-exclusion fencing. Wildlife Research 36:410–421.

Conover, M. R. 2002. Resolving wildlife conflicts: the science of wildlife damage management. CRC Press, Boca Raton, Florida, USA.

Côté, S. D., T. P. Rooney, J. Tremblay, C. Dussault, and D. M. Waller. 2004. Ecological impacts of deer overabundance. Annual Review of Ecology, Evolution, and Systematics 34:113–147.

Craven, S. R, and S. E. Hygnstrom. 1994. Deer. Pages D25–D40 in S. E. Hygnstrom, R. M. Timm, and G. E. Larson, editors. Prevention and control of wildlife damage. University of Nebraska–Lincoln, Lincoln, USA.

Curtis, P. D., M. J. Farigone, and M. E. Richmond. 1994. Preventing deer damage with barrier, electrical, and behavioral fencing systems. Proceedings of the Vertebrate Pest Control Conference 16:223–227.

D'Angelo, G. J., J. G. D'Angelo, G. R. Gallagher, D. A. Osborn, K. V. Miller, and R. J. Warren. 2007. Evaluation of wildlife warning reflectors for altering white-tailed deer behavior along roadways. Wildlife Society Bulletin 34:1175–1183.

deCalesta, D. S., and M. G. Cropsey. 1978. Field test of a coyote-proof fence. Wildlife Society Bulletin 6:256–259.

DeNicola, A. J., K. C. VerCauteren, P. D. Curtis, and S. E. Hygnstrom. 2000. Managing white-tailed deer in suburban environments. Cornell University Cooperative Extension, Ithaca, New York, USA.

DeVault, T. L., J. L. Belant, B. F. Blackwell, and T. W. Seamans. 2011. Interspecific variation in wildlife hazards to aircraft: implications for airport wildlife management. Wildlife Society Bulletin 35:394–402.

DeVault, T. L., J. E. Kubel, D. J. Glista, and O. E. Rhodes. 2008. Mammalian hazards at small airports in Indiana: impact of perimeter fencing. Human–Wildlife Conflicts 2:240–247.

Dillon, J. L. 1977. The analysis of response in crop and livestock production. Second edition. Pergamon, Oxford, United Kingdom.

Ditchkoff, S. S., and B. C. West. 2007. Ecology and management of feral hogs. Human–Wildlife Conflicts 1:149–151.

Dolbeer, R. A. 1988. Current status and potential of lethal means of reducing bird damage in agriculture. Pages 474–483 in H. Ouellet, editor. Acta XIX Congressus Internationalis Ornithologici. University of Ottawa, Ottawa, Ontario, Canada.

Dolbeer, R. A., S. E. Wright, and E. C. Cleary. 2000. Ranking the hazard level of wildlife species to aviation. Wildlife Society Bulletin 28:372–378.

Dolbeer, R. A., S. E. Wright, J. R. Weller, and M. J. Begier. 2011. Wildlife strikes to civil aircraft in the United States 1990–2009. Serial Report Number 16. U.S. Department of Transportation, Federal Aviation Administration, Office of Airport Safety and Standards, Washington, D.C., USA.

Duffy, B., B. McBratney, B. Holland, and D. Colvert. 1988. Fences. U.S. Department of the Interior, Bureau of Land Management, U.S. Department of Agriculture Forest Service, Technology and Development Program, Washington, D.C., USA.

Dyer, M. I., and P. Ward. 1977. Management of pest situations. Pages 267–300 in J. Pinowski, S. C. Kendeigh, editors. Granivorous birds in ecosystems. Cambridge University Press, New York, New York, USA.

Falk, N. W., H. B. Graves, and E. D. Bellis. 1978. Highway right-of-way fences as deer deterrents. Journal of Wildlife Management 42:646–650.

Feldhamer, G. A., J. E. Gates, D. M. Harman, A. J. Loranger, and K. R. Dixon. 1986. Effects of interstate highway fencing on white-tailed deer activity. Journal of Wildlife Management 50:497–503.

Fitzwater, W. D. 1972. Barrier fencing in wildlife management. Proceedings of the Vertebrate Pest Conference 5:49–55.

Flyger, V., and T. Thoerig. 1962. Crop damage caused by Maryland deer. Proceedings of the Southeastern Association of Game and Fish Commissioners 16:45–52.

Goddard, P. J., R. W. Summers, A. J. MacDonald, C. Murray, and A. R. Fawcett. 2001. Behavioural responses of red deer to fences of five different designs. Applied Animal Behaviour Science 73:289–298.

Hawkins, R. E., W. D. Klimstra, and D. C. Autry. 1971. Dispersal of deer from Crab Orchard National Wildlife Refuge. Journal of Wildlife Management 35:216–220.

Honda, T., Y. Miyagawa, H. Ueda, and M. Inoue. 2009. Effectiveness of newly-designed electric fences in reducing crop damage by medium and large mammals. Mammal Study 34:13–17.

Hygnstrom, S. E., and S. R. Craven. 1988. Electric fences and commercial repellents for reducing deer damage in cornfields. Wildlife Society Bulletin 16:291–296.

Isleib, J. 1995. Deer exclusion efforts to reduce crop damage in Michigan and northeast Wisconsin. Proceedings of the Great Plains Wildlife Damage Control Workshop 12:63–69.

Kaneene, J. B., C. S. Bruning-Fann, L. M. Granger, R. Miller, and B. A. Porter-Spalding. 2002. Environmental and farm management associated with tuberculosis on cattle farms in northeastern Michigan. Journal of the American Veterinary Medical Association 221:837–842.

Lavelle, M. J., K. C. VerCauteren, T. J. Hefley, G. E. Phillips, S. E. Hygnstrom, D. B. Long, J. W. Fischer, S. R. Swafford, and T. A. Campbell. 2011. Evaluation of fences for containing feral swine under simulated depopulation conditions. Journal of Wildlife Management 75:1200–1208.

Leblond, M., C. Dussault, J. Ouellet, M. Poulin, R. Courtois, and J. Fortin. 2007. Electric fencing as a measure to reduce moose–vehicle collisions. Journal of Wildlife Management 71:1695–1703.

Lehnert, M. E., and J. A. Bissonette. 1997. Effectiveness of highway crosswalk structures at reducing deer–vehicle collisions. Wildlife Society Bulletin 25:809–818.

Marsh, R. E., A. E. Koehler, and T. P. Salmon. 1990. Exclusionary methods and materials to protect plants from pest mammals—a review. Proceedings of the Vertebrate Pest Conference 14:174–180.

Matschke, G. H., D. S. deCalesta, and J. D. Harder. 1984. Crop damage and control. Pages 647–654 in L. K. Halls, editor. White-tailed deer: ecology and management. Stackpole Books, Harrisburg, Pennsylvania, USA.

McAninch, J. B. 1986. Recent advances in repellents and fencing to deter deer damage. Proceedings of the New England Fruit Meetings 86:31–39.

McKillop, I. G., and R. M. Sibly. 1988. Animal behaviour at electric fences and the implications for management. Mammal Review 18:91–103.

Miller, F. L., C. J. Jonkel, and G. D. Tessier. 1972. Group cohesion and leadership response by barren-ground caribou to man-made barriers. Artic 25:193–202.

Palmer, W. L., J. M. Payne, R. G. Wingard, and J. L. George. 1985. A practical fence to reduce deer damage. Wildlife Society Bulletin 13:240–245.

Peterson, M. N., R. R. Lopez, N. J. Silvy, C. B. Owen, P. A. Frank, and A. W. Braden. 2003. Evaluation of deer-exclusion grates in urban areas. Wildlife Society Bulletin 31:1198–1204.

Poole, D. W., G. Western, and I. G. McKillop. 2004. The effects of fence voltage and the type of conducting wire on the efficacy of an electric fence to exclude badgers (Meles meles). Crop Protection 23:27–33.

Porter, W. F. 1983. A baited electric fence for controlling deer damage to orchard seedlings. Wildlife Society Bulletin 11:325–329.

Reed, G. J., R. J. Warren, K. V. Miller, G. R. Gallagher, and S. A. Valitzski. 2007. Development and evaluation of devices designed to minimize deer–vehicle collisions. Daniel B. Warnell School of Forestry and Natural Resources, University of Georgia, Athens, USA.

Reidy, M. M., T. A. Campbell, and D. G. Hewitt. 2010. Evaluation of electric fencing to inhibit feral pig movements. Journal of Wildlife Management 72:1012–1018.

Sauer, P. R. 1984. Physical characteristics. Pages 73–90 in L. K. Halls, editor. White-tailed deer: ecology and management. Stackpole Books, Harrisburg, Pennsylvania, USA.

Seamans, T. W. 2001. A review of deer control devices intended for use on airports. Proceedings of the 3rd joint annual meeting. Bird Strike Committee–USA/Canada, 27–30 August 2001, Calgary, Alberta, Canada.

Seamans, T. W., and D. A. Helon. 2008. Evaluation of an electrified mat as a white-tailed deer (Odocoileus virginianus) barrier. International Journal of Pest Management 54:89–94.

Seamans, T. W., and K. C. VerCauteren. 2006. Evaluation of ElectroBraid™ as a white-tailed deer barrier. Wildlife Society Bulletin 34:8–15.

Smith, D. 1983. Deer control using 5 strand vertical fence. Proceedings of the Eastern Wildlife Damage Control Conference 1:97–98.

Steger, R. E. 1988. Consider using electric powered fences for controlling animal damage. Proceedings of the Great Plains Wildlife Damage Control Workshop 8:215–216.

Stull, D. W., W. D. Gulsby, J. A. Martin, G. J. D'Angelo, G. R. Gallagher, D. A. Osborn, R. J. Warren, and K. V. Miller. 2011. Comparison of fencing designs for excluding deer from roadways. Human–Wildlife Interactions 5:47–57.

Thompson, B. C. 1978. Fence-crossing behavior exhibited by coyotes. Wildlife Society Bulletin 6:14–17.

Tierson, W. C. 1969. Controlling deer use of forest vegetation with electric fences. Journal of Wildlife Management 33:922–926.

Van Noord, J. R. 2000. Bambi be gone! Flying Safety 56:18–21.

VerCauteren, K. C., S. E. Hygnstrom, R. M. Timm, R. M. Corrigan, J. G. Beller, L. L. Bitney, M. C. Brumm, D. Meyer, D. R. Virchow, and R. W. Wills. 2002. Development of a model to assess rodent control in swine facilities. Pages 59–64 in L. Clark, J. Hone, J. Shivik, R. Watkins, K. C. VerCauteren, and J. Yoder, editors. Human conflicts with wildlife: economic considerations. U.S. Department of Agriculture, Wildlife Services, National Wildlife Research Center, Fort Collins, Colorado, USA.

VerCauteren, K. C., M. J. Lavelle, and S. E. Hygnstrom. 2006a. Fences and deer-damage management: a review of designs and efficacy. Journal of Wildlife Management 34:191–200.

VerCauteren, K. C., M. J. Lavelle, and S. E. Hygnstrom. 2006b. A simulation model for determining cost-effectiveness of fences for reducing deer damage. Wildlife Society Bulletin 34:16–22.

VerCauteren, K. C., N. W. Seward, M. J. Lavelle, J. W. Fischer, and G. E. Phillips. 2009. Deer guard and bump gates for excluding white-tailed deer from fenced resources. Human–Wildlife Conflicts 3:145–153.

VerCauteren, K. C., T. R. VanDeelen, M. J. Lavelle, and W. H. Hall. 2010. Assessment of abilities of white-tailed deer to jump fences. Journal of Wildlife Management 74:1378–1381.

Ward, A. L. 1982. Mule deer behavior in relation to fencing and underpasses on Interstate 80 in Wyoming. Transportation Research Record 859:8–13.

Wilson, C. J. 1993. Badger damage to growing oats and an assessment of electric fencing as a means of its reduction. Journal of Zoology 231:668–675.

6

Wildlife Translocation as a Management Alternative at Airports

Paul D. Curtis
Jonathon D. Cepek
Rebecca Mihalco
Thomas W. Seamans
Scott R. Craven

Wildlife in urban settings may be a welcome sight for many, but negative interactions between people and various wild species are increasing (Conover et al. 1995, Conover 2002). Wildlife populations are commonly managed in part to reduce these conflicts, particularly in high-risk areas such as roadways and airports (Conover 2002). However, the public often opposes lethal control or management methods perceived as causing harm to nuisance animals (Reiter et al. 1999, Conover 2002, Treves et al. 2006), and attitudes vary considerably depending on the particular wildlife species involved (Kretser et al. 2009). Consequently, a variety of nonlethal management approaches are typically integrated with limited lethal control (Conover 2002).

Translocation, the transport and release of wild animals from one location to another (Nielsen 1988), is an example of a fairly recent adaptation to wildlife damage management methods. Griffith et al. (1989) provided an overview of translocation as a general wildlife conservation method. Translocation has been demonstrated as an important technique for stocking game species and furbearers, reintroducing extirpated species, and enhancing threatened or endangered species. The black bear (*Ursus americanus*) is probably the carnivore most frequently translocated to re-establish extirpated populations (Smith and Clark 1994, Linnell et al. 1997). Based on a survey of 81 wildlife agencies and organizations (1973–1986), Griffith et al. (1989) determined that 90% of all translocations were of native game species and were deemed successful 86% of the time. In contrast, translocations of threatened species were successful only 46% of the time.

Translocation also has been used to remove problem carnivores in the hope that the negative experience will prevent the animal from returning to the conflict site, or that the individual will stay near the release area, where the potential for future conflicts is low (Rogers 1988, Gunther 1994, Linnell et al. 1997). The translocation of felids has been a common management method to reduce livestock depredations, especially in Africa (Linnell et al. 1997). Holevinski et al. (2006) reported that few (seven of 80, or 8.8%) Canada geese (*Branta canadensis*) translocated ~150 km (93 miles) from urban areas returned to their original capture site during the six months following banding. Most geese were harvested <50 km from their release site during the fall hunting season following summer banding. In contrast, hazed Canada geese repeatedly returned to airport environments because they were apparently habituated to nonlethal control methods (York et al. 2000).

Translocation is a viable management tool to reestablish raptor breeding populations, including Seychelles kestrel (*Falco araea*; Watson 1989) and osprey (*Pandion haliaetus*; Martell et al. 2000; see additional references in Cade and Temple 1994), but it has generally received equivocal reviews when applied to damage management scenarios (Linnell et al. 1997, Thirgood et al. 2000, Watson and Thirgood 2001). Vacant territories of golden eagles (*Aquila chrysaetos*) translocated to reduce predation on livestock were quickly taken over by other eagles, and 14 of 16 eagles eventually returned to their capture sites (Phillips et al. 1991). Despite a paucity of data, translocation of raptors is deemed an effective and socially acceptable management tool to reduce the abundance of these birds at airports as well as the frequency of bird–aircraft collisions (i.e., bird strikes; see Schafer et al. 2002).

Because both airport biologists and the public seem to support raptor translocation despite a lack of data, there is a need to realistically assess the effectiveness of this method. We first briefly review the legal and ecological concerns (across wildlife species) associated with animal translocation and the reasons why this management tool is used at airports. We then discuss management data on raptor translocations from airports and how these data can be used to assess relative costs and benefits versus alternative management options.

Legal Concerns

In their national survey examining translocation of nuisance wildlife, Craven and Nosek (1992) reported that 47 states allowed the translocation of animals from the site of capture. Some states had species-specific restrictions, often against species identified as carriers of rabies. Most states reported that euthanasia was the preferred management alternative for handling urban nuisance animals, although 41 states reported that euthanasia was not mandatory for any species. Twenty-eight states required a state-issued permit, license, or permission from the appropriate wildlife agency to translocate wildlife. Fourteen states allowed anyone with nuisance wildlife to capture and remove the problem animals. Similarly, La Vine et al. (1996) found that fish and wildlife agencies in 33 states allowed property owners to translocate animals causing damage or conflicts, and eight states allowed any species to be translocated; 13 states had regulations prohibiting translocation of threatened or endangered species. Wildlife agencies in 45 states allowed property owners to euthanize animals causing damage or conflicts, and 42 states restricted species that could be handled by private personnel.

With regard to capture and translocation of raptors, the overriding legal issue is their protection under the Migratory Bird Treaty Act of 1918 (see the Digest of Federal Resource Laws of the U.S. Fish and Wildlife Service [USFWS]; http://www.fws.gov/laws/lawsdigest/migtrea .html). A USFWS migratory bird depredation permit is necessary for capture and translocation or lethal removal of protected migratory birds (http://www.fws.gov/ migratorybirds/mbpermits.html), and state depredation permits might be required in addition to the federal permit. For airports dealing with management of bald eagles (*Haliaeetus leucocepalus*) to reduce strike hazards, an eagle depredation permit from the USFWS is also required. Although bald eagles were removed from the Endangered Species List in 2007, they remain protected under the Bald and Golden Eagle Protection Act of 1940 (http://www.fws.gov/laws/lawsdigest/baldegl.html).

Ecological Concerns

Survival of released animals is often lower than that for established, wild individuals. Rosatte and MacInnes (1989) reported a 50% mortality rate for translocated raccoons within three months after release. In addition to high mortality rates for translocated animals, there are long-distance movements and increased risk of disease transmission (Wright 1978). Bendel and Therres (1994) reported that only 55% of 20 translocated Delmarva fox squirrels (*Sciurus niger*) survived 90 days postrelease. Transmission of infectious disease to resident wildlife (Rosatte and MacInnes 1989) is also a risk that might not be readily noticed or discernible at the time of translocation. There is extensive literature on raptor site fidelity to breeding areas (e.g., Janes 1984, Jenkins and Jackman 1993, Rosenfield and Bielefeldt 1996; see also winter area site fidelity in Garrison and Bloom 1993, Hinnebusch et al. 2010) and homing abilities (Boshoff and Vernon 1988, Latta et al. 2005, Linthicum et al. 2007), factors that could limit successful translocation. Craven et al. (1998) suggested the following guidelines for successful wildlife translocation: (1) proper selection of a release site, including landowner permission and suitable habitat; (2) consideration of season and weather conditions, time of day, and distances from capture sites at time of release; and (3) adherence to recommendations for health certification or quarantine for certain species.

Translocation to Reduce Bird Strikes

Raptor–Aircraft Strikes

Survival of translocated animals, and risks to the wildlife community at the release site, are clearly important. However, one must also consider the probability of death associated with the animal's use of airport habitats if not translocated, as well as hazards posed to human health and safety. Blackwell and Wright (2006) found that most aircraft strikes (63%) with red-tailed hawks (*Buteo jamaicensis*) occurred while the plane was on the ground, and 84% of strikes occurred below 30.5 m (100 feet) above ground level, all within the airport environment. In addition, from 1990 through 2009, the U.S. Department of Transportation Federal Aviation Administration (FAA) National Wildlife Strike Database (FAA 2011) showed that raptors (including vultures and owls) were responsible for 5,724 reported strikes, resulting in almost $56 million in reported economic losses (Dolbeer et al. 2011). Most strike-related damage to civil aircraft involved bald eagles ($14,402,681), vultures ($9,312,759), and red-tailed hawks ($6,709,526; Dolbeer et al. 2011). These loss estimates are likely conservative, as the reporting rate was estimated at only 20% from 1990 through 1994 and 39% from 2004 through 2008, and only 14% of these reports indicated damage (Dolbeer et al. 2011).

More recently, DeVault et al. (2011) ranked species and groups according to their relative hazard to aircraft when struck in the airport environment (i.e., 152 m [≤500 feet] above ground level). The authors used a composite rank reflecting the percentage of total strikes (for that species or species group) that caused any level of damage to the aircraft, the percentage of total strikes that caused substantial damage to the aircraft (for definitions of aircraft damage categories, see Dolbeer et al. 2000), and the percentage of total strikes that caused an effect on flight. Of the 66 bird species or groups examined, five species of raptors and turkey vultures (*Cathartes aura*) ranked among the top 20 for relative hazard score. The management of raptors and vultures is a high priority for biologists charged with reducing wildlife hazards at airports.

Management Example: Raptors at Ohio Airports

At civilian airports in Ohio, USA, 3,162 bird strikes were reported to the FAA (1990–2009), with hawks, owls, and vultures involved in 290 strikes (FAA 2011). American kestrels (*F. sparverius*) accounted for 46% of the raptor strikes, red-tailed hawks were responsible for

23%, and unknown hawks and short-eared owls (*Asio flammeus*) added 9% each. Peregrine falcons (*F. peregrinus*) and turkey vultures contributed 3% each. The remaining 7% consisted of several species of hawks and owls (FAA 2011). In 2004, the U.S. Department of Agriculture (USDA) Wildlife Services (WS) Ohio program obtained authorization from the USFWS to translocate raptors. This decision provided enhanced opportunities for nonlethal management of raptors using airports in Ohio. Lethal control of raptors was used when there were no other reasonable options, or when it was necessary to remove a bird that was an immediate and direct hazard to aircraft operations. Additionally, WS developed a peregrine falcon translocation plan because of two aircraft strikes with juvenile falcons in 2004. Because peregrines were listed as an endangered species in Ohio during 2004 (currently peregrine falcons are listed as a state-threatened species in Ohio), WS did not pursue permission to lethally remove them.

To further reduce hazards while conserving Ohio's state-listed raptors, and based on perceived public support in favor of nonlethal raptor management, WS developed a raptor and owl relocation plan in collaboration with the Ohio Department of Natural Resources Division of Wildlife (USDA 2009). Under this agreement, translocation of raptors would be used only when repeated harassment attempts failed to resolve the problem. During 2009, WS biologists captured and translocated 33 American kestrels and 31 red-tailed hawks from a single Ohio airport (USDA 2010; Fig. 6.1).

In 2010, managers translocated an additional 25 kestrels and 46 red-tailed hawks, with translocation distances ranging from 72 to 120 km (45 to 75 miles). All 135 birds captured at Ohio airports during this time period were marked with USFWS leg bands to evaluate potential recovery rates. Recovery rates were low for these banded raptors (see also McIlveen et al. 1992/93, Schafer et al. 2002). Five banded red-tailed hawks were recovered within the original airport environment in 2009 and 2010. Airport personnel shot two hawks, and three were recaptured and euthanized (one was found injured as the result of a suspected aircraft collision).

The efforts in Ohio reflect a nationwide trend for WS. From 2008 through 2010, WS biologists translocated 606 red-tailed hawks from 19 airports (313 hatching-year birds, 293 after-hatching-year birds; L. Schafer, WS, unpublished data). Overall, the confirmed

Fig. 6.1. Red-tailed hawk captured within a Swedish goshawk trap positioned near a runway at an airport in Ohio, USA. Rock pigeons (*Columba livia*), protected by the cage, served as lures. Photo credit: U.S. Department of Agriculture, Wildlife Services

return rate was 6% (39 of 606). The confirmed return rate based on distance translocated was similar for both juvenile and adult banded hawks. Peak months for capture of after-hatching-year red-tailed hawks were February and March, whereas hatching-year hawks were more likely to be caught and relocated during September and October.

Summary

Raptor translocation from airports shows promise relative to hazard reduction, but the cost-effectiveness of such programs has not been clearly demonstrated. The cost-effectiveness of this management approach should be assessed relative to continued integration of other nonlethal management strategies (e.g., reducing habitat and food resources), as well as to lethal control as a last measure. Important variables to be considered when evaluating all management approaches for raptors include (1) staff time, (2) equipment needs, (3) documentation of return rates for raptor species (including sex, age, location, season, and distance of translocation), (4) relative reduction in strike rates, and (5) estimates of survival of translocated birds versus mortality rates for individuals remaining in airport environments.

LITERATURE CITED

Bendel, P. R., and G. D. Therres. 1994. Movements, site fidelity and survival of Delmarva fox squirrels following translocation. American Midland Naturalist 132:227–233.

Blackwell, B. F., and S. E. Wright. 2006. Collisions of red-tailed hawks (*Buteo jamaicensis*), turkey vultures (*Cathartes aura*), and black vultures (*Coragyps atratus*) with aircraft: implications for bird strike reduction. Journal of Raptor Research 40:76–80.

Boshoff, A. F., and C. J. Vernon. 1988. The translocation and homing ability of problem eagles. South African Journal of Wildlife Research 18:38–40.

Cade, T. J., and S. A. Temple. 1994. Management of threatened bird species: evaluation of the hands on approach. Ibis 137:S161–S172.

Conover, M. R. 2002. Resolving human–wildlife conflicts. CRC Press, Boca Raton, Florida, USA.

Conover, M. R., W. C. Pitt, K. K. Kessler, T. J. DuBow, and W. A. Sanborn. 1995. Review of human injuries, illnesses, and economic losses caused by wildlife in the United States. Wildlife Society Bulletin 23:407–414.

Craven, S. R., T. Barnes, and G. Kania. 1998. Toward a professional position on the translocation of problem wildlife. Wildlife Society Bulletin 26:171–177.

Craven, S. R., and J. A. Nosek. 1992. Final report to the NPCA: summary of a survey on translocation of suburban wildlife. University of Wisconsin, Department of Wildlife Ecology, Madison, USA.

DeVault, T. L., J. L. Belant, B. F. Blackwell, and T. W. Seamans. 2011. Interspecific variation in wildlife hazards to aircraft: implications for airport wildlife management. Wildlife Society Bulletin 35:394–402.

Dolbeer, R. A., S. E. Wright, and E. C. Cleary. 2000. Ranking the hazard level of wildlife species to aviation. Wildlife Society Bulletin 28:372–378.

Dolbeer, R. A., S. E. Wright, J. R. Weller, and M. J. Begier. 2011. Wildlife strikes to civil aircraft in the United States, 1990–2009. Serial Report No. 16. U.S. Department of Transportation, Federal Aviation Administration, Washington, D.C., USA.

FAA. Federal Aviation Administration. 2011. FAA wildlife strike database. http://wildlife-mitigation.tc.faa.gov/wildlife/default.aspx.

Garrison, B. A., and P. H. Bloom. 1993. Natal origins and winter site fidelity of rough-legged hawks wintering in California. Journal of Raptor Research 27:116–118.

Griffith, B., J. M. Scott, J. W. Carpenter, and C. Reed. 1989. Translocation as a species conservation tool: status and strategy. Science 245:477–480.

Gunther, K. A. 1994. Bear management in Yellowstone National Park. Proceedings of the International Conference on Bear Research and Management 9:549–561.

Hinnebusch, D. M., J. F. Therrien, M. A. Valiquette, B. Robertson, S. Robertson, and K. L. Bildstein. 2010. Survival, site fidelity, and population trends of American kestrels wintering in southwestern Florida. Wilson Journal of Ornithology 122:475–483.

Holevinski, R. A., R. A. Malecki, and P. D. Curtis. 2006. Can hunting of translocated nuisance Canada geese reduce local conflicts? Wildlife Society Bulletin 34:845–849.

Janes, S. W. 1984. Fidelity to breeding territory in a population of red-tailed hawks. Condor 86:200–203.

Jenkins, J. M., and R. E. Jackman. 1993. Mate and nest site fidelity in a resident population of bald eagles. Condor 95:1053–1056.

Kretser, H. E., P. D. Curtis, J. D. Francis, R. J. Pendall, and B. A. Knuth. 2009. Factors affecting perceptions of human–wildlife interactions in residential areas of northern New York and implications for conservation. Human Dimensions of Wildlife 14:102–118.

Latta, B. C., D. D. Driscoll, J. L. Linthicum, R. E. Jackman, and G. Doney. 2005. Capture and translocation of golden eagles from the California Channel Islands to mitigate depredation of endemic island foxes. Pages 341–350 in D. K. Garcelon and C. A. Schwemm, editors. Proceedings of the sixth California islands symposium. National Park Service Technical Publication CHIS-05-01. Institute for Wildlife Studies, Arcata, California, USA.

La Vine, V. P., M. J. Reeff, J. A. Dicamillo, and G. S. Kania. 1996. The status of nuisance wildlife damage control in the United States. Proceedings of the Vertebrate Pest Conference 17:8–12.

Linnell, J. D. C., R. Aanes, and J. E. Swenson. 1997. Translocation of carnivores as a method for managing problem animals: a review. Biodiversity and Conservation 6:1245–1257.

Linthicum, J., R. E. Jackman, and B. C. Latta. 2007. Annual migrations of bald eagles to and from California. Journal of Raptor Research 41:106–112.

Martell, M. S., J. V. Englund, and H. B. Tordoff. 2000. An urban osprey population established by translocation. Journal of Raptor Research 36:91–96.

McIlveen, W. D., M. Wernaart, and D. Brewer. 1992/93. Update on the raptor relocation program at Pearson International Airport, 1989–1993. Ontario Bird Banding 25/26:64–70.

Nielsen, L. 1988. Definitions, considerations, and guidelines for translocation of wild animals. Pages 12–51 in L. Nielsen and R. D. Brown, editors. Translocation of wild animals. Wisconsin Humane Society, Milwaukee, USA.

Phillips, R. L., J. L. Cummings, and J. D. Berry. 1991. Responses of breeding golden eagles to relocation. Wildlife Society Bulletin 19:430–434.

Reiter, D. K., M. W. Brunson, and R. H. Schmidt. 1999. Public attitudes toward wildlife damage management and policy. Wildlife Society Bulletin 27:746–758.

Rogers, L. L. 1988. Homing tendencies of large mammals: a review. Pages 76–92 in L. D. Nielsen and R. D. Brown, editors. Translocation of wild animals. Wisconsin Humane Society, Milwaukee, USA.

Rosatte, R. C., and C. D. MacInnes. 1989. Relocation of city raccoons. Proceedings of the Great Plains Wildlife Damage Control Conference 9:87–92.

Rosenfield, R. N., and J. Bielefeldt. 1996. Lifetime nesting area fidelity in male Cooper's hawks in Wisconsin. Condor 98:165–167.

Schafer, L. M., J. L. Cummings, J. A. Yunger, and K. E. Gustad. 2002. Evaluation of raptor translocation at O'Hare International Airport, Chicago, Illinois. Final report to U.S. Department of Transportation. Federal Aviation Administration, Washington, D.C., USA.

Smith, K. G., and J. D. Clark. 1994. Black bears in Arkansas: characteristics of a successful translocation. Journal of Mammalogy 75:309–320.

Thirgood, S., S. Redpath, I. Newton, and P. Hudson. 2000. Raptors and red grouse: conservation conflicts and management solutions. Conservation Biology 14:95–104.

Treves, A., R. B. Wallace, L. Naughton-Treves, and A. Morales. 2006. Co-managing human–wildlife conflicts: a review. Human Dimensions of Wildlife 11:383–396.

USDA. U.S. Department of Agriculture. 2009. Raptor and owl relocation plan for Ohio. Animal and Plant Health Inspection Service, Wildlife Services, Cleveland, Ohio, USA.

USDA. U.S. Department of Agriculture. 2010. Annual report of activities for Cleveland Hopkins International Airport and Burke Lakefront Airport, Cleveland, Ohio. Animal and Plant Health Inspection Service, Wildlife Services, Cleveland, Ohio, USA.

Watson, J. 1989. Successful translocation of the endemic Seychelles kestrel (Falco araea) to Praslin. Pages 363–367 in B. U. Meyburg and R. D. Chancellor, editors. Raptors in the modern world. World Working Group on Birds of Prey, Berlin, Germany.

Watson, M., and S. Thirgood. 2001. Could translocation aid hen harrier conservation in the UK? Animal Conservation 4:37–43.

Wright, G. A. 1978. Dispersal and survival of translocated raccoons in Kentucky. Proceedings of the Southeastern Association of Fish and Wildlife Agencies 31:285–294.

York, D. L., J. L. Cummings, R. M. Engeman, and K. L. Wedemeyer. 2000. Hazing and movements of Canada geese near Elmendorf Air Force Base in Anchorage, Alaska. International Biodeterioration and Biodegradation 45:103–110.

7

Richard A. Dolbeer
Alan B. Franklin

Population Management to Reduce the Risk of Wildlife–Aircraft Collisions

Four basic control strategies mitigate the risks to aviation caused by wildlife at airports: (1) aircraft flight schedule modification (primarily at military airbases) and enhancement of aircraft visibility to avoid interactions with wildlife (e.g., Blackwell et al. 2009b, 2012); (2) habitat modification and elimination of food, water, and cover that attract wildlife (Cleary and Dolbeer 2005, Blackwell et al. 2009a; Chapters 5, 8–10); (3) repellent and harassment techniques to disperse wildlife (Cleary and Dolbeer 2005; Chapters 2–4); and (4) wildlife population management (e.g., Dolbeer 1998). As discussed throughout this book, successful efforts to mitigate the risk of wildlife–aircraft strikes at airports usually involve programs that attempt to integrate these strategies. This chapter focuses on wildlife population management.

In general, wildlife population reduction by killing or through reproductive control at or in the vicinity of an airport is the last option deployed after all other actions have been considered or implemented. However, management of a wildlife hazard situation at an airport may require killing an individual animal, or require that a local population of a problem species be reduced by lethal or reproductive means until, if feasible, a long-term, nonlethal solution can be implemented (e.g., erecting a deer-proof fence, relocating a nearby gull [Laridae] nesting colony; see Chapters 5–6). In addition, lethal removal of a few individuals sometimes reinforces nonlethal frightening techniques (Baxter and Allan 2008). Recurrent lethal control is often necessary as part of an integrated Wildlife Hazard Management Plan (WHMP) for an airport (Cleary and Dolbeer 2005, Baxter 2008).

Most wildlife species that frequent airport environments are protected by some combination of federal, state, and local laws, often requiring permits before any action can be taken to capture or kill animals or to control their reproduction. Ninety percent of the birds struck by civil aircraft in the USA are species federally protected by the Migratory Bird Treaty Act (Dolbeer et al. 2012). Permits require justification of why the removal is needed, the numbers to be removed by species, and the methods used to remove and dispose of the animals. In addition, management of wildlife populations often generates public interest, which airports must acknowledge and address. The following steps should be taken to justify population reduction through lethal or reproductive control and to minimize adverse public reaction to a program involving killing wildlife:

- Document that the wildlife species is an economic, safety, or health threat.
- Justify why nonlethal options alone are not adequate to solve the problem.
- Assess the impact that the lethal or reproductive control will have on local and regional populations of the species (i.e., is the action likely to result in a significant reduction in numbers of the species at the local or regional level?).
- Assure that the methods are appropriate (i.e., legal, safe, effective, and humane) and specific for the targeted wildlife species.

- Document the number of animals killed or treated by species.
- Document the effectiveness of population management actions in mitigating the problem (e.g., reduction in numbers observed at airports and in wildlife strikes).
- Recommend steps to be taken, if any are feasible, to reduce the need for population management actions in the future.
- Issue timely reports, preferably annually, that summarize the items listed above. Transparency increases public acceptance and allows for more effective adaptive management strategies.

Three critical types of information are needed for airports to justify lethal or reproductive control programs to regulatory agencies and the public before implementing these programs. First, the hazard level and the risk posed by the wildlife species must be documented (Dolbeer and Wright 2009). Lethal control may be warranted at a particular airport for species such as Canada geese (*Branta canadensis*) or white-tailed deer (*Odocoileus virginianus*) that have a high hazard level (i.e., >50% of strikes with aircraft result in damage; Dolbeer et al. 2012) and that pose a high risk (i.e., the species have been documented through observations or strike events to frequent the airport; see also Biondi et al. 2011, DeVault et al. 2011). In contrast, at the same airport it may be inappropriate to request a permit for lethal control for a species such as American kestrel (*Falco sparverius*) with a relatively low hazard level (<2% of strikes cause damage) and that is infrequently observed.

An understanding of the local and regional population status and dynamics of the problem species is also needed before developing a management plan. Population data from local surveys, breeding bird surveys, Christmas bird counts, and other sources can be integrated with reproductive and survival rates to develop simple population models for the species of concern (Dolbeer 1998, Runge et al. 2009). These models can predict the immediate impact that lethal or reproductive control programs will have on local or regional populations and project how populations will respond to these management actions (e.g., Blackwell et al. 2003, Runge et al. 2009). Such models provide a scientific foundation to guide management actions

and to provide a level of objectivity in the emotional debates that often arise when proposals are made to kill or reduce reproductive rates of wildlife (Dolbeer 1998).

Finally, airports must monitor the population level of the targeted species, as well as the number of strikes and associated damage (DeVault et al. 2011) caused by that species before and after implementing the population management plan. Monitoring allows for documentation of the effects that management actions have on the population and, most importantly, on the number of strikes.

These three types of information would be ideally integrated into regional strategic plans that encompass all airports within a specified area, allowing for more efficient permitting, implementation, and monitoring of target wildlife species. An emphasis on regional, rather than national, strategies takes into account that problem wildlife species in one area may not necessarily be problems in another area. In addition, the incorporation of adaptive management into regional strategic plans would allow for more efficient "learning while doing." Adaptive management is a formal, structured process that allows for flexible decision making in the face of uncertain outcomes from management practices and natural variability (Williams et al. 2007). Successfully used in regional management of natural resources (e.g., Weinstein et al. 1996), this approach has direct application to management of wildlife populations at airports.

Primer of Population Dynamics

Any consideration of management of wildlife populations by airport biologists, particularly lethal management, should be grounded in a basic understanding of wildlife population dynamics from spatial and temporal perspectives. In particular, effects of population demography (age and sex ratios, reproductive rates) and seasonal habitat and foraging requirements will influence how populations use airport environments. For most widely distributed wildlife populations, airports represent relatively small management units that may be used differently depending on the season. As such, wildlife habitats within airport perimeters probably do not sustain distinct population segments, but environments outside airport perimeters bolster these populations. Airports generally represent microcosms within a larger landscape, and effective management of wildlife within these microcosms depends on the species, characteristics of their population dynamics, their habitats, and other spatial and temporal factors affecting their populations.

Wildlife populations occur at a variety of spatial scales, ranging from small, isolated populations to continent-wide populations. These populations can also vary temporally (e.g., daily, seasonally, or annually) at any given location. Rates of population growth (λ) for wildlife populations depend on several species-specific demographic components, such as annual survival, reproductive output, immigration, and emigration. In terms of measurable quantities, λ can be expressed as:

$$\lambda = \phi + f \,,$$

where ϕ is apparent survival for older age classes (a function of survival and emigration) and f is recruitment (a function of reproductive output and immigration; Nichols et al. 2000). In turn, these demographic characteristics are dependent on spatial factors such as habitat suitability and quality, as well as temporal factors such as seasonal weather conditions. Under the concept of r and K selection (Stearns 1976, Boyce 1984), there exists a continuum of life history strategies relevant to population dynamics, where r-selected species mature early and have high reproductive output and low adult survival (low ϕ and high f in the above equation), and where K-selected species mature late and have low reproductive output and high adult survival (high ϕ and low f in the above equation). This range of different life-history strategies will affect the success of methods used to manage populations (see below).

One key, underlying factor controlling population dynamics is habitat (e.g., Pulliam and Danielson 1991). Habitat is a species-specific concept; each species has unique habitat requirements. In terms of populations, habitat quality is a key concept and can be defined as the "ability of the environment to provide conditions appropriate for individual and population persistence" (Hall et al. 1997). Habitat quality is linked inextricably with population performance from small to large scales. Habitat quality governs larger-scale metapopulation processes such as source–sink dynamics, where population sources may reside in areas of high-quality habitat that then contribute individuals to areas of low-quality habitat (sinks) through recruitment (primarily immi-

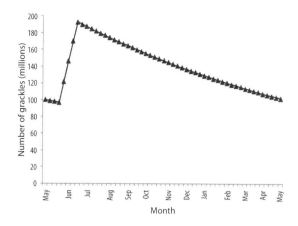

Fig. 7.1. Simulated annual cycle of the common grackle population in eastern North America, demonstrating the dynamic nature of wildlife populations. Adapted from Dolbeer (1998)

gration; Pulliam 1988, Pulliam and Danielson 1991, Runge et al. 2006; but see also Doncaster et al. 1997).

Also, unlike human populations, wildlife populations often exhibit dramatic within-year (annual) cycles in numbers. Most wildlife species have a narrow season of births followed by fledging/weaning, which introduces a large pulse of young animals into the population each year. This pulse of young animals occurs in summer for most species at the middle to high latitudes typical of Europe and North America. The magnitude of the annual population cycle is related to the age-specific reproductive rate of the species. Species such as snowshoe hares (*Lepus americanus*; Dolbeer and Clarke 1975) and red-winged blackbirds (*Agelaius phoeniceus*; Dolbeer et al. 1976) have pronounced annual cycles because females are sexually mature at one year old and are capable of producing several young each year. The common grackle (*Quiscalus quiscula*, a species with a similar life history as the red-winged blackbird) population in the eastern USA is estimated to be about 100 million at the start of the nesting season in April. By June, when young have fledged (a mean of about two per female one year and older), the population has almost doubled to about 200 million. For the long-term population to remain stable, natural mortality must eliminate about 100 million grackles between June and the following April for the population to begin the next annual cycle at 100 million birds (Dolbeer et al. 1997*b*, Dolbeer 1998; Fig. 7.1).

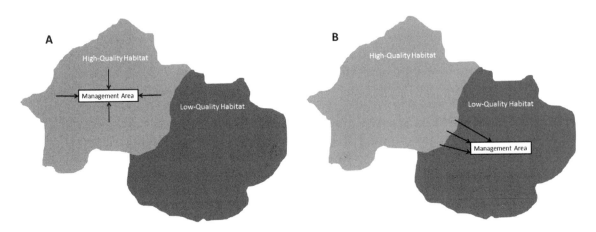

Fig. 7.2. Example of how surrounding habitat quality can affect management of overabundant populations. Management is conducted on a small scale in (A) a source population and (B) an adjacent sink population. Arrows indicate recruitment of new individuals into the management area.

In years when natural factors (e.g., inclement weather, disease) increase mortality or decrease reproduction, intraspecific competition may be reduced, with wildlife populations typically responding with increased survival or reproduction. Conversely, if natural factors result in an exceptional year of successful reproduction or low mortality, subsequent increased competition for food and habitat typically reduces reproduction or survival. These compensatory factors (Caughley 1977) dampen fluctuations in annual population levels and can stabilize the population in the long term. Exceptions occur with fundamental changes in habitat quality or mortality/reproductive factors. For example, the dramatic increase in the double-crested cormorant (*Phalacrocorax auritus*) population in the Great Lakes in the 1980s and 90s resulted from the combination of increased reproduction (elimination of chlorinated hydrocarbon pesticides) and decreased mortality (protection by Migratory Bird Treaty Act and enhanced food supply through the introduction of large-scale fish farming in the southern USA; Hatch 1995). Many other large bird species in North America and Europe exhibited similar increases in populations from 1980 through 1999 because of fundamental changes in carrying capacity (Dolbeer and Eschenfelder 2003).

Understanding the factors contributing to population fluctuations is especially relevant to the management of overabundant populations. Managing such populations in small areas may achieve temporary reductions, but these reductions may fail over the long term if the spatial and temporal scales and the factors governing dynamics at those scales are not considered. In a hypothetical simple source–sink system where high-quality habitat represents a source of individuals and low-quality habitat represents a sink, management of an overabundant population at a small scale will likely require repeated removals over multiple years because (1) removed individuals within the management unit in the source population will be replaced by recruitment from the surrounding population (Fig. 7.2A), and (2) individuals removed from the management unit embedded within the sink population will be replaced (possibly at a slower rate) by recruits from the adjacent source population (Fig. 7.2B). Under this scenario, one viable management option may be collaboration between airport managers and biologists with local municipalities and land owners to reduce desired habitat to less desired habitat for those species being managed (e.g., Blackwell et al. 2009*a*). Because the risk to aviation safety must be mitigated at airports, the removal of wildlife that disperse into the air operations area (AOA), even when habitat management and harassment programs are in place to discourage such dispersal, is often an ongoing part of the airport's WHMP (see case studies below).

Another aspect of population dynamics, one applicable to management of wildlife populations that pose hazards to aviation, involves the concept and practice of reproductive control to manage overabundant wildlife populations that are causing conflicts with humans.

Because the urbanized public generally advocates non-lethal means of managing problem populations of wildlife, there has been increased interest in the development of reproductive control strategies for wildlife species (Fagerstone et al. 2010). However, the modeling of population responses to various levels of lethal and reproductive control clearly demonstrate that for almost all species, lethal control is more efficient in reducing populations than reproductive control (Dolbeer et al. 1988, Dolbeer 1998, Blackwell et al. 2002). The exceptions are some small rodent and bird species with high reproductive rates and low survival rates (Dolbeer 1998)—species that pose little hazard to aviation (Dolbeer and Wright 2009). That reproductive control (e.g., oiling eggs in nests of gulls [*Larus* spp.] or Canada geese) may take several years to reduce the target population size makes this approach unacceptable for solving immediate risks posed by wildlife to aviation.

Population Management to Reduce Wildlife Strikes at U.S. Airports: Case Studies

There are numerous situations in which lethal control has been implemented to resolve human conflicts with wildlife at airports. In 2011, U.S. Department of Agriculture biologists used some level of lethal wildlife control at 314 civil and military airports in the USA as part of integrated management programs (Begier and Dolbeer 2012). Lethal control also has been used frequently in other (nonaviation) situations, such as agriculture, to reduce human–wildlife conflicts (Dolbeer 1986, Bedard et al. 1995, Dolbeer et al. 1997b). The following three case studies from airports demonstrate the utility of lethal control as part of integrated management programs.

Gulls at John F. Kennedy International Airport

Gull–aircraft collisions have long been a serious problem at John F. Kennedy International Airport (JFK), New York, New York, USA. Gulls, of which 60% were laughing gulls (*L. atricilla*), caused 86% of bird strikes from 1988 through 1990, averaging 261 strikes per year. Laughing gulls are present from May through September in association with a nesting colony at Jamaica

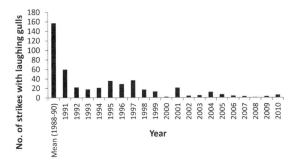

Fig. 7.3. Number of strike events involving laughing gulls at John F. Kennedy International Airport, New York, New York, USA (1988–2010). A shooting program was implemented in 1991.

Bay Wildlife Refuge, which is adjacent to the airport. Although the airport implemented numerous nonlethal actions to reduce gull presence at the airport in the 1980s, the number of strikes increased as the nesting gull population increased in the adjacent wildlife refuge (Dolbeer et al. 1993).

As an alternative approach to reduce strikes in 1991 (and continuing through 2011), biologists started a population management program in which managers stationed on JFK airport boundaries shot gulls flying over the airport from May through August. As a result of the shooting program, the number of strikes with laughing gulls was reduced to 38% of 1988–1990 levels in 1991 (the first year) and to 1–5% of 1988–1990 levels in 2008–2011 (Washburn et al. 2009, R. A. Dolbeer, unpublished data). Strikes by the three other gull species were reduced to 10–52% of preshooting levels over the same time periods. In 1991 and 1992, about 14,000 and 12,000 laughing gulls, respectively, were killed; this number declined to about 2,000–6,000 gulls in subsequent years (Washburn et al. 2009, R. A. Dolbeer, unpublished data). The laughing gull colony in Jamaica Bay has declined 73%, from 7,629 nests in 1990 to 2,040 nests in 2011 (Dolbeer et al. 1997a, Washburn and Tyson 2011). That the colony size declined by 73% from 1990 to 2011 while the annual strike rate of laughing gulls declined by over 95% (2008–2011; Fig. 7.3) indicated that many laughing gulls altered flight patterns and avoided the airport in response to shooting (Dolbeer et al. 2003). Although the shooting program has reduced the local population of gulls flying over JFK (Fig. 7.4), the regional population (>300,000 birds),

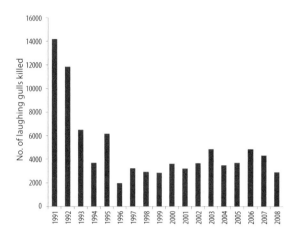

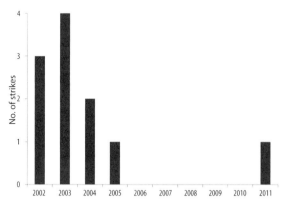

Fig. 7.4. Number of laughing gulls killed by shotgun at John F. Kennedy International Airport, New York, New York, USA (1991–2008).

Fig. 7.5. Number of Canada goose strikes with aircraft <152 m (500 feet) above ground level at LaGuardia Airport, New York, New York, USA. Years comprise the twelve months from July through July (e.g., 2002 is July 2002 through June 2003); 2011 represents July–December only. From June 2004 through June 2011, 1,456 geese were removed from Rikers Island.

as predicted by modeling, has not been negatively impacted (Dolbeer 1998, Dolbeer et al. 2003). This study demonstrated that shooting can significantly reduce gull–aircraft collisions at an airport by both reducing the local population (but not the regional population) and altering flight patterns of surviving gulls.

Canada Geese near LaGuardia Airport

The resident (nonmigratory) population of Canada geese increased dramatically in North America from about 0.25 million in 1970 to 3.47 million in 2010 (Dolbeer and Seubert 2011), posing a substantial hazard to aircraft (Dolbeer and Eschenfelder 2003). In the 1990s, a portion of the growing population of resident Canada geese in New York City began using Rikers Island as a gathering site during the molting season (June–July). Rikers Island is located in the East River, about 0.5 km (0.3 miles) from LaGuardia Airport (LGA), New York, New York, USA. During the two-year period from July 2002 to June 2004, seven Canada goose strikes were recorded at LGA (all at <152 m (500 feet) above ground level; Fig. 7.5). These strikes included a passenger aircraft departing LGA in September 2003 that hit at least five Canada geese, causing an uncontained failure in one engine and requiring an emergency landing at JFK, 18 km (11 miles) away (National Transportation Safety Board 2004).

As a result of these strikes, a population manage-

ment program was initiated at Rikers Island in June 2004 in which 518 resident geese, representing over 90% of the geese using the island, were rounded up during the molt (when they are flightless) and euthanized. In the seven subsequent years, the number of geese removed from the island steadily declined to 55 in 2011 (Fig. 7.6). The number of strikes at LGA involving Canada geese at <152 m above ground level (and thus in the airport environment) also declined in the aftermath of the management program (Fig. 7.5). Compared to the seven strikes recorded in the two years before the first removal at Rikers (June 2004), there have been only four strikes in the subsequent seven years. Two of those four strikes occurred in August–September 2004, less than three months after the first removal; there have been only two strikes in the subsequent seven years (October 2004 to December 2011). This focused population management program resulted in a major reduction in the local population of Canada geese near the airport and the number of strikes by this high-risk species. This program, involving the removal of 1,456 geese from 2004 to 2011, has had no impact on the regional population. The metropolitan area of New York City currently contains 15,000–20,000 resident Canada geese (B. Swift, New York State Department of Environmental Conservation, personal communication; Collins and Humberg 2011).

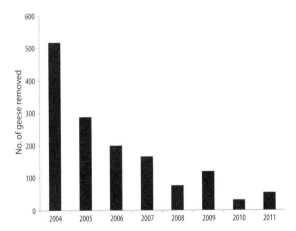

Fig. 7.6. Number of Canada geese removed from Rikers Island, New York, New York, USA, during the molt period (late June) from 2004 through 2011. Each year, >90% of the geese on the island were captured and euthanized.

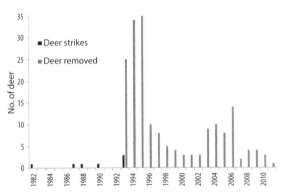

Fig. 7.7. Number of aircraft collisions with white-tailed deer at Chicago O'Hare International Airport, Chicago, Illinois, USA (1982–2011). In December 1993, 25 deer were shot and removed at night. From 1994 through 2011, 160 additional deer were killed. There has not been a deer strike at O'Hare since November 1993.

Deer at Chicago O'Hare International Airport

Deer at airports pose one of the highest risks of any wildlife species to departing and arriving aircraft (Wright et al. 1998, Dolbeer and Wright 2009, Biondi et al. 2011, DeVault et al. 2011, Dolbeer et al. 2012). Deer-proof fencing is the best long-term approach for excluding deer from AOAs; Chapter 5). However, larger airports may require >15 km (9 miles) of fencing to secure the AOA, often traversing uneven ground with numerous gates and culverts. Even with good fencing, it is not uncommon for deer to enter AOAs, especially in areas with high deer populations (DeVault et al. 2008). From 1990 through 2010, civil aircraft struck about 1,000 deer (*Odocoileus* spp.) at airports in the USA (Dolbeer et al. 2012).

In 1993, aircraft struck three deer at Chicago O'Hare International Airport (ORD), Chicago, Illinois, USA, prompting emergency action. In December 1993, sharpshooters removed 25 deer from the AOA at night, followed by the removal of 34, 35, 10, and 8 deer in 1994, 1995, 1996, and 1997, respectively. By 1998, the perimeter fence had been improved substantially to exclude deer, but deer still occasionally entered the AOA. Up to 14 deer were removed per year from 1998 to 2011 (Fig. 7.7). When appropriate, deer removed from the airport were processed and donated to charitable organizations.

The combination of lethal control starting in December 1993 and improved fencing resulted in no deer strikes at ORD in the subsequent 19 years from November 1993 through 2011. The overall deer population density in the Chicago area has not been estimated but is considered high (Etter et al. 2002); in 2005, Cook County (where ORD and Chicago are located) had about 1,000 deer–automobile collisions, the highest of any county in Illinois (Flood 2008). The overall deer population in Illinois is about 800,000, with over 150,000 harvested by hunters annually (Channick 2010); clearly, the removal program at ORD has not adversely affected local or regional deer abundance.

Summary

Lethal management of wildlife on and near airport properties is often an essential component of integrated management actions to mitigate the risk of wildlife–aircraft strikes. Despite the potentially catastrophic consequences of wildlife strikes, however, lethal management often evokes contention from the public. Management decisions involving population reduction must therefore be based on (1) an understanding of the factors affecting wildlife population dynamics, (2) the integration of lethal management with nonlethal methods, and (3) observational data before, during, and after implementation. These observational data (numbers killed, population levels, and number of strikes with aircraft) are critical to determine the impact of lethal management actions on each wildlife species' population and on the mitigation of risk to aircraft using the airport. This information should be compiled into periodic reports (typically annually) that are made available to the public.

LITERATURE CITED

Baxter, A. 2008. The impact of lethal control as a reinforcement technique when deploying IBSC best practice standards on an aerodrome. Proceedings of the 28th annual IBSC conference. International Bird Strike Committee. 24–28 November 2008, Brasilia, Brazil.

Baxter, A. T., and J. R. Allan. 2008. Use of lethal control to reduce habituation to blank rounds by scavenging birds. Journal of Wildlife Management 72:1653–1657.

Bedard, J., A. Nadeau, and M. Lepage. 1995. Double-crested cormorant culling in the St. Lawrence River estuary. Colonial Waterbirds 18:78–85.

Begier, M. J., and R. A. Dolbeer. 2012. Protecting the flying public and minimizing economic losses within the aviation industry: technical, operational, and research assistance provided by USDA-APHIS-Wildlife Services to reduce wildlife hazards to aviation, fiscal year 2011. U.S. Department of Agriculture, Animal and Plant Health Inspection Service, Wildlife Services, Washington, D.C., USA.

Biondi, K. M., J. L. Belant, J. A. Martin, T. L. DeVault, and G. Wang. 2011. White-tailed deer incidents with U.S. civil aircraft. Wildlife Society Bulletin 35:303–309.

Blackwell, B. F., T. L. DeVault, E. Fernández-Juricic, and R. A. Dolbeer. 2009a. Wildlife collisions with aircraft: a missing component of land-use planning on and near airports? Landscape and Urban Planning 93:1–9.

Blackwell, B. F., T. L. DeVault, T. W. Seamans, S. L. Lima, P. Baumhardt, and E. Fernández-Juricic. 2012. Exploiting avian vision with aircraft lighting to reduce bird strikes. Journal of Applied Ecology 49:758–766.

Blackwell, B. F., E. Fernández-Juricic, T. W. Seamans, and T. Dolans. 2009b. Avian visual configuration and behavioural response to object approach. Animal Behaviour 77:673–684.

Blackwell, B. F., E. Huszar, G. Linz, and R. A. Dolbeer. 2003. Lethal control of red-winged blackbirds to manage damage to sunflower: an economic evaluation. Journal of Wildlife Management 67:818–828.

Blackwell, B. F., M. A. Stapanian, and D. V. Weseloh. 2002. Evaluating dynamics of the double-crested cormorant population on Lake Ontario. Wildlife Society Bulletin 30:345–353.

Boyce, M. S. 1984. Restitution of r- and K-selection as a model of density-dependent natural selection. Annual Review of Ecology and Systematics 15:427–447.

Caughley, G. 1977. Analysis of vertebrate populations. John Wiley and Sons, Brisbane, Queensland, Australia.

Channick, R. 2010. Deer population going strong, despite 2 decades of culling programs. Chicago Tribune. http://articles.chicagotribune.com/2010-11-24/news/ct-x-n-deer-culling-20101124_1_deer-management-deer-population-deer-herds.

Cleary, E. C., and R. A. Dolbeer. 2005. Wildlife hazard management at airports: a manual for airport personnel. Second edition. Federal Aviation Administration, Office of Airport Safety and Standards, Washington, D.C., USA.

Collins, R., and L. Humberg. 2011. Summary: New York City Canada goose removals in 2011. U.S. Department of Agriculture, Wildlife Services, Albany, New York, USA.

DeVault, T. L., J. L. Belant, B. F. Blackbird, and T. W. Seamans. 2011. Interspecific variation in wildlife hazards to aircraft: implications for airport wildlife management. Wildlife Society Bulletin 35:394–402.

DeVault, T. L., J. E. Kubel, D. J. Glista, and O. E. Rhodes. 2008. Mammalian hazards at small airports in Indiana: impact of perimeter fencing. Human–Wildlife Conflicts 2:240–247.

Dolbeer, R. A. 1986. Current status and potential of lethal means for reducing bird damage in agriculture. Acta XIX Congressus Internationalis Ornithologici 1:474–483.

Dolbeer, R. A. 1998. Population dynamics: the foundation of wildlife damage management for the 21st century. Proceedings of the Vertebrate Pest Conference 18:2–11.

Dolbeer, R. A., J. L. Belant, and G. E. Bernhardt. 1997a. Aerial photography techniques to estimate populations of laughing gulls in Jamaica Bay, New York, 1992–1995. Colonial Waterbirds 20:8–13.

Dolbeer, R. A., J. L. Belant, and J. Sillings. 1993. Shooting gulls reduces strikes with aircraft at John F. Kennedy International Airport. Wildlife Society Bulletin 21:442–450.

Dolbeer, R. A., R. B. Chipman, A. L. Gosser, and S. C. Barras. 2003. Does shooting alter flight patterns of gulls: case study at John F. Kennedy International Airport. Proceedings of the International Bird Strike Committee 26:547–564.

Dolbeer, R. A., and W. Clark. 1975. Population dynamics of the snowshoe hare in the central Rocky Mountains. Journal of Wildlife Management 39:535–549.

Dolbeer, R. A., and P. Eschenfelder. 2003. Amplified bird-strike risks related to population increases of large birds in North America. Proceedings of the International Bird Strike Committee 26:49–67.

Dolbeer, R. A., L. A. Fiedler, and H. Rasheed. 1988. Management of giant fruit bat and rat populations in the Maldive Islands, Indian Ocean. Proceedings of the Vertebrate Pest Conference 13:112–118.

Dolbeer, R. A., C. R. Ingram, and J. L. Seubert. 1976. Modeling as a management tool for assessing the impact of blackbird control measures. Proceedings of the Vertebrate Pest Conference 7:35–45.

Dolbeer, R. A., D. F. Mott, and J. L. Belant. 1997b. Blackbirds and starlings killed at winter roosts from PA-14 applications: implications for regional population management. Proceedings of the Eastern Wildlife Damage Management Conference 7:77–86.

Dolbeer, R. A., and J. L. Seubert. 2011. Canada goose populations and strikes with civil aircraft, 1990–2010: encouraging trends for the aviation industry. U.S. Department of Agriculture, Wildlife Services, Airport Wildlife Hazards Program, Washington, D.C., USA.

Dolbeer, R. A., and S. E. Wright. 2009. Safety management systems: how useful will the FAA National Wildlife Strike Database be? Human–Wildlife Conflicts 3:167–178.

Dolbeer, R. A., S. E. Wright, J. Weller, and M. J. Beiger. 2012. Wildlife strikes to civil aircraft in the United States, 1990–2010. Serial Report No. 17. U.S. Department of Transportation, Federal Aviation Administration, Office of Airport Safety and Standards, Washington, D.C., USA.

Doncaster, C. P., J. Clobert, B. Doligez, L. Gustafsson, and E. Danchin. 1997. Balanced dispersal between spatially varying local populations: an alternative to the source–sink model. American Naturalist 150:425–445.

Etter, D. R., K. M. Hollis, T. R. Van Deelen, D. R. Ludwig, J. E. Chelsvig, C. L. Anchor, and R. E. Warner. 2002. Survival and movements of white-tailed deer in suburban Chicago, Illinois. Journal of Wildlife Management 66:500–510.

Fagerstone, K. A., L. A. Miller, G. Killian, and C. A. Yoder. 2010. Review of issues concerning the use of reproductive inhibitors, with particular emphasis on resolving human–wildlife conflicts in North America. Integrative Zoology 1:15–30.

Flood, S. 2008. Final report of the joint task force on deer population control as required by House Joint Resolution 65. Illinois General Assembly, Illinois Department of Natural Resources, Deer Task Force, Springfield, USA.

Hall, L. S., P. R. Krausman, and M. L. Morrison. 1997. The habitat concept and a plea for standard terminology. Wildlife Society Bulletin 25:173–182.

Hatch, J. J. 1995. Changing populations of double-crested cormorants. Colonial Waterbirds 18:8–24.

National Transportation Safety Board. 2004. Accident report NYC03FA190 for Fokker 100 (N1450A) departing La-Guardia Airport, New York, New York, 4 September 2003. Washington, D.C., USA.

Nichols, J. D., J. E. Hines, J.-D. Lebreton, and R. Pradel. 2000. Estimation of contributions to population growth: a reverse-time capture-recapture approach. Ecology 81:3362–3376.

Pulliam, H. R. 1988. Sources, sinks, and population regulation. American Naturalist 132:652–661.

Pulliam, H. R., and B. J. Danielson. 1991. Sources, sinks, and habitat selection: a landscape perspective on population dynamics. American Naturalist 137:S50–S66.

Runge, J. P., M. C. Runge, and J. D. Nichols. 2006. The role of local populations within a landscape context: defining and classifying sources and sinks. American Naturalist 167:925–938.

Runge, M. C., J. R. Sauer, M. L. Avery, B. F. Blackwell, and M. D. Koneff. 2009. Assessing allowable take of migratory birds: black vultures in Virginia. Journal of Wildlife Management 73:556–565.

Stearns, S. C. 1976. Life-history tactics: a review of the ideas. Quarterly Review of Biology 51:3–47.

Washburn, B. E., B. N. Haslun, M. S. Lowney, and S. M. Tennis. 2009. Shooting gulls to reduce strikes with aircraft at John F. Kennedy International Airport, 1991–2008. Special report for the Port Authority of New York and New Jersey, John F. Kennedy International Airport. U.S. Department of Agriculture, Wildlife Services, National Wildlife Research Center, Sandusky, Ohio, USA.

Washburn, B. E., and L. A. Tyson. 2011. Laughing gull nest population in Jamaica Bay, New York, 1992–2011. Special report for the Port Authority of New York and New Jersey, John F. Kennedy International Airport. U.S. Department of Agriculture, Wildlife Services, National Wildlife Research Center, Sandusky, Ohio, USA.

Weinstein, M., J. Balletto, J. Teal, and D. Ludwig. 1996. Success criteria and adaptive management for a large-scale wetland restoration project. Wetlands Ecology and Management 4:111–127.

Williams, B. K., R. C. Szaro, and C. D. Shapiro. 2007. Adaptive management: the U.S. Department of Interior technical guide. U.S. Department of the Interior, Adaptive Management Working Group, Washington, D.C., USA.

Wright, S. E., R. A. Dolbeer, and A. J. Montoney. 1998. Deer on airports: an accident waiting to happen. Proceedings of the Vertebrate Pest Conference 18:90–95.

PART II · MANAGING RESOURCES

TRAVIS L. DEVAULT
BRIAN E. WASHBURN

8

Identification and Management of Wildlife Food Resources at Airports

Wildlife use airport habitats for a variety of reasons, including breeding, raising young, resting, taking refuge from predators, and locating sources of water. But the chief motivation for most individuals to encroach on airports is food. Depending on the specific habitat types present and habitat management strategies employed, airports can harbor large numbers of small mammals, insects, earthworms, and palatable vegetation that attract many species hazardous to aircraft. Often the best way to reduce populations of hazardous wildlife at airports is to determine which sources of food are being used, and then remove or modify those foods to make them less attractive (Washburn et al. 2011). Fortunately, the science of wildlife ecology and management has a long and productive history of research on wildlife food habits and foraging strategies, and the applied nature of most food habit studies conducted in airport environments facilitates straightforward specialization of investigational techniques. In this chapter we (1) discuss in more detail food resources as a primary motivation for wildlife use of airport properties, (2) consider some established principles of wildlife food habits and foraging strategies that affect airport wildlife management, (3) review techniques used to investigate wildlife food habits and identify those most useful for airports, (4) discuss methods for eliminating or modifying some preferred foods at airports, and (5) briefly consider future research needs.

Although we focus our discussion on birds (>97% of all wildlife–aircraft strikes involve birds), white-tailed deer (*Odocoileus virginianus*; Biondi et al. 2011) and other mammals (Dolbeer et al. 2010) present significant hazards at some airports. Even so, deer and many other mammals can be managed effectively with exclusion techniques (Chapter 5). For airports without adequate fencing, the food habits of deer, coyotes (*Canis latrans*), and other hazardous mammals should be considered when developing wildlife hazard management protocols. For example, even though few birds regularly feed on soybeans (Sterner et al. 1984, Krapu et al. 2004), deer are major consumers of soybean plants (Humberg et al. 2007), and thus soybean cultivation should be discouraged at and near airports without adequate fencing.

Food: A Primary Motivation for Wildlife Use of Airports

Why are so many wildlife species attracted to airports? There are many reasons. Although they can contain a variety of habitat types (Blackwell et al. 2009, DeVault et al. 2009), airports are usually characterized by wide-open spaces relatively free from human activity. DeVault et al. (2012) calculated that airports in the USA certificated by the Federal Aviation Administration (FAA; see Appendix) contain an average of 297 ha of grassland. Airports also have stormwater treatment facilities and other water bodies that can attract hazardous wildlife (Chapter 9).

If one considers the three basic needs of wildlife—food, water, and shelter—wildlife can readily obtain all three at an airport. On closer investigation, however, water and shelter may be less problematic overall and easier for airport biologists to manage than food resources. Water bodies certainly do attract waterfowl and other hazardous wildlife to airports, and at times offer considerable management challenges. Even so, water attractants are usually identified easily, and substantial progress has been made in recent years in the design and management of water bodies at airports to deter use by hazardous wildlife (Chapter 9). As for shelter, the overall homogeneity of airport lands relative to off-airport areas helps to limit refuge and loafing areas for some types of hazardous wildlife. Biologists can identify and remove mammal dens and raptor nests, and close hangars and other airport buildings to deny access to rock pigeons (*Columba livia*), European starlings (*Sturnus vulgaris*), and other birds closely associated with humans. But because wildlife food resources are so abundant and take so many different forms, it is difficult—if not impossible—to remove them completely. Even at airports employing full-time wildlife biologists, wildlife consistently forage on airport properties.

An examination of the FAA's National Wildlife Strike Database (Dolbeer et al. 2010) indicates that hazardous wildlife use airports primarily for foraging, as opposed to nesting, loafing, and other activities. Blackwell et al. (2013) reviewed database records from 1990 to 2008 and determined that of the nine grassland-associated bird species that caused the most damaging strikes to aircraft, only killdeer (*Charadrius vociferous*) commonly nest in airport grasslands. The remaining bird species—Canada goose (*Branta canadensis*) red-tailed hawk (*Buteo jamaicensis*), and European starling—use airport grasslands primarily for foraging on grasses, small mammals, and insects, respectively. These data suggest that proper management of food resources at airports could help reduce strike risk by reducing wildlife foraging in critical areas.

Other studies have indicated that food resources are primary determinants of bird movements and spatial ecology (i.e., where and how birds choose to spend their time). Rolando (2002:53) reviewed factors affecting home range characteristics and determined that "food availability is the primary determinant of home range ecology in birds and all other factors are secondary." Further evidence for the importance of food resources

on bird movement behaviors is illustrated by black vultures (*Coragyps atratus*) and turkey vultures (*Cathartes aura*), common North American scavengers (Kirk and Mossman 1998, Buckley 1999) that are particularly hazardous to aircraft (Dolbeer et al. 2000, DeVault et al. 2011). Black and turkey vultures adjust their home range characteristics, movement patterns, and even flight behaviors based on the local nature of carrion resources. Coleman and Fraser (1987, 1989) studied black and turkey vultures in an agricultural region of Pennsylvania and Maryland, USA, and found that they relied heavily on carrion from domestic (farm) animals, a relatively predictable and constant source of food. Conversely, DeVault et al. (2004, 2005) and Kelly et al. (2007) investigated movement behaviors and food habits of both species in a heavily forested environment in South Carolina, USA, and found that those vultures relied almost exclusively on carrion from wild animals, a more ephemeral and unpredictable source of food. DeVault et al. (2004, 2005) also determined that vultures in their heavily forested study area had much larger home ranges (~100% larger) and spent a greater percentage of daylight hours in flight (approximately two to five times more) than their counterparts in Pennsylvania and Maryland (Coleman and Fraser 1989). DeVault and colleagues concluded that differences in habitat structure—and, by extension, food resources—presented a more challenging foraging environment to vultures in the heavily forested region in South Carolina, reflected by the substantial differences in their spatial ecology across the two environments. This plasticity in vulture behavior across their ranges underscores the importance of food resources on bird movements and demonstrates how the manipulation of food resources can potentially influence wildlife activity patterns at airports.

Principles of Wildlife Food Habits and Foraging Strategies

Research on wildlife food habits has a long history in wildlife research and management, and there is a well-developed literature on theory and application. In this section we consider a few of those topics that we believe are especially important for airport investigations. Readers interested in a general discussion of methods for investigating wildlife food habits and subsequent management strategies are encouraged to see Litvaitis (2000) and McDonald et al. (2005).

Use, Selection, and Preference

Although the terms "use," "selection," and "preference" are often used interchangeably, there are important differences among them (Johnson 1980, Litvaitis 2000, McDonald et al. 2005). Use is nothing more than the consumption of a particular food, whereas selection occurs when an animal chooses a certain food item when others are more readily available; that is, disproportionally to its availability. Preference for a food is independent from its availability; preference can be inferred only when foods are equally available. Questions of food preference are generally not addressed in the airport context, because necessary study designs to investigate such questions (e.g., cafeteria-style experiments or enclosures where resources are carefully controlled) are usually not practical. Therefore food selection is most often the variable of interest at airports. Mere use of a food does not necessarily imply that eliminating that food will influence behavior of the species consuming it, because that species might simply switch to an equally desirable food (Litvaitis et al. 1994) available at the airport. Identification of food resources that hazardous wildlife select, however, is an important component of effective management.

Abundance versus Availability

Many studies on wildlife food habits have measured and reported food abundance, rather than availability, because of the difficulty in measuring true availability (Litvaitis 2000, McDonald et al. 2005). However, it is availability—the proportion of a food resource that is accessible—that influences food selection by wildlife (Johnson 1980). Buckley and McCarthy (1994) studied laughing gulls (*Larus atricilla*) at John F. Kennedy International Airport (JFK), New York, New York, USA, and found that gulls fed on adult Oriental beetles (*Anomala orientalis*) only in shortgrass areas, even though the same beetles were equally abundant (but much more difficult to capture, and thus less available) in nearby tallgrass areas. Another example of abundance versus availability that is particularly applicable to airport wildlife management concerns the issue of small

mammals as prey for raptors (i.e., hawks, eagles, and falcons), which present substantial hazards to aircraft (DeVault et al. 2011). Several researchers have investigated habitat use of a variety of raptor species as it relates to prey densities across habitat types; these authors consistently reported that prey availability (a function of both prey density and vulnerability to predation), rather than abundance, most strongly correlates with habitat use (e.g., Wakely 1978, Baker and Brooks 1981*a,b*, Bechard 1982, Preston 1990, Beier and Drennan 1997). To consider one example in particular, Baker and Brooks (1981*a*) studied the distribution of red-tailed hawks and rough-legged hawks (*B. lagopus*) at Toronto International Airport, Toronto, Ontario, Canada. In their study, both hawk species were more numerous in shortgrass areas than on straw fields or old fields (both of which had taller vegetation and less bare ground), despite lower densities of their most common prey, meadow voles (*Microtus pennsylvanicus*), in shortgrass areas. Baker and Brooks (1981*a*) concluded that meadow voles were more vulnerable to predation in shortgrass areas, which led to increased raptor use of those fields.

Dietary Breadth

In normal circumstances, many animals use fewer types of foods than they are physiologically capable of consuming. But during food shortages, animals often increase the diversity of their food habits (Litvaitis 2000), and some species regularly use a surprisingly wide variety of foods. We propose that, in general, wildlife have more diverse diets than is commonly believed (see also Polis 1991). For example, snakes are often thought of exclusively as predators of small animals, but wild snakes regularly consume carrion (including road-killed frogs that are peeled from the road surface; DeVault and Krochmal 2002) and have been known to consume cooked spareribs (Savidge 1988), slaughtered pig (Heinrich and Studenroth 1996), and canned dog food (Parker and McCallum 2010). Further evidence concerning the dietary breadth of wildlife comes from studies using remote cameras to study predation of bird nests. Such studies have demonstrated that various squirrel species (usually considered herbivores) can be major nest predators (Sieving and Willson 1998, Williams and Wood 2002, Grant et al. 2006), and have documented white-tailed deer (also herbivores) eating grassland bird nestlings (Pietz and Granfors 2000). Given the ability and occasional motivation of various species to consume "unusual" foods, managers must keep an open mind regarding wildlife food habits when investigating and managing food resources at airports. As an example, Bernhardt et al. (2009) found that tree swallows (*Tachycineta bicolor*), which generally feed on flying insects, fed almost exclusively on fruits from bayberry bushes (*Myrica pensylvanica*) at JFK. Removal of bayberry bushes resulted in a 75% reduction in tree swallow–aircraft strikes at the airport.

Constraints on Optimal Foraging

Theory suggests that animals forage in a way that maximizes energy intake and minimizes energy expenditure (e.g., prey capture and handling time; MacArthur and Pianka 1966, Shoener 1971). Decades of studies on optimal foraging theory (see Shoener 1986) have been helpful in developing our understanding of foraging behavior, including food selection. However, optimal foraging theory is a simplification (Litvaitis 2000). In reality, many other factors influence foraging behavior, including nutritional content, intra- and interspecific competition, body condition, sex and age class, environmental conditions, and (most notably in the current context) risk of predation (Lima and Dill 1990, Lima 1998). Such constraints on optimal foraging behavior are important to recognize, because it may be possible to use these constraints in the context of airport wildlife management. Blackwell et al. (2013) discuss how vegetation could potentially be managed to enhance the perceived risk of predation and thus reduce frequency of foraging on airport grasslands by some bird species.

Techniques for Investigating Wildlife Food Habits at Airports

Accurate determination of food selection by wildlife at airports usually requires collecting food samples from regurgitated pellets or gastrointestinal tracts, although direct observation of foraging behaviors and feeding site surveys is possible in some circumstances. We discuss techniques to investigate wildlife food habits, concentrating on those most useful for airports. More

exhaustive treatment of food habit analysis techniques is available in Rosenberg and Cooper (1990), Litvaitis et al. (1994), and McDonald et al. (2005).

We emphasize that priority for study, as well as for subsequent management actions, should be placed on those species that are most hazardous to aircraft; that is, those most likely to cause damage or to have a negative effect on flight when struck (Dolbeer et al. 2000, Dolbeer and Wright 2009, DeVault et al. 2011). Such information is important because alteration of a food resource to decrease airport use by one species might inadvertently (and unavoidably) create an attractant for another. A priori knowledge of relative hazard level, as well as established wildlife–habitat relationships, helps to inform priorities for study and management.

Although choosing the specific techniques to study food habits generally depends on the question being addressed (see McDonald et al. 2005), in the airport context the questions are usually fairly consistent: what foods do hazardous wildlife at this airport select, and how can I subsequently remove or modify those foods so that they are no longer selected by that species? Airport investigations of food habits are somewhat unique in that the investigator is most interested in what the animal eats within certain administrative boundaries. Diet composition of focal individuals outside airport property (assuming that the airport does not constitute the entire home range) is somewhat less important, because management of food resources outside the airport boundary is often impractical or impossible. Even so, food selection for an individual can occur at scales larger than the airport property (especially for birds; Martin et al. 2011), and the portion of the home range occupied by the airport could contain anywhere from all to none of the food regularly consumed by that individual. Common examples include Canada geese feeding on airport turfgrass but nesting in an adjacent wetland, or gulls feeding in a nearby landfill but loafing on the airport pavement. When possible, one must understand the food selection of hazardous airport wildlife in a larger context. This knowledge can help discern the contribution to an animal's diet of food resources found in airport and off-airport habitats, as well as those specific to a particular airport.

The sample size necessary for accurate representation of food habits will vary depending on season, variability in diet across individuals, and dietary breadth, and for this reason it is difficult to determine before study initiation (Rosenberg and Cooper 1990, Litvaitis et al. 1994). Fortunately, when using individuals killed during control activities or birds struck by aircraft, sample size is generally not an issue—one simply uses all the birds available, or at a minimum continues analysis until no more unique information is added to the data set (Rosenberg and Cooper 1990). However, the location of collection can heavily influence study results. Washburn et al. (2011) compared stomach contents from European starlings collected at JFK on airport grasslands and near the shoreline to birds struck by aircraft, and found that only birds collected on grasslands had diets similar to those of struck birds. Because food habits of birds involved in actual strikes with aircraft provide the most relevant data to airport investigations, such samples should be used whenever possible. Because the availability of aircraft-struck birds is limited at most airports, however, it is often necessary to obtain diet samples by other means (see below). Care should be taken to ensure that samples are representative of individuals most vulnerable to aircraft strikes. Mangers must also consider the most appropriate temporal and demographic sampling scheme for collections. Within species, food needs often change seasonally (Williams and Jackson 1981, Fischl and Caccamise 1987, Bernhardt et al. 2010) and across age and sex classes (Litvaitis et al. 1994).

Several techniques can be used to obtain dietary samples or to observe foraging activities at airports, each with advantages and disadvantages (Table 8.1). The most common and preferred technique, as inferred above, is the use of gastrointestinal tracts from birds struck by aircraft or collected during wildlife control activities (Chapter 7). Stomach contents and bird crops can provide a multitude of diet information and can be analyzed by sex, age, and reproductive class (Fig. 8.1). The study of gastrointestinal tracts is also favored because the samples are readily provided—it is not necessary to collect animals specifically for study. Even though such samples are conveniently obtained, however, they might be limited in number and, in the case of samples collected during control activities, might not accurately reflect the diets of individuals actually struck by aircraft. The analysis of regurgitated pellets (birds) and feces (mammals) is also commonly employed. These techniques are inexpensive, minimally invasive, and can yield a great deal of information. Unfortunately, analysis of these samples often suffers from bias due to differential digestibility of various food types (Litvaitis et al. 1994), and usually does not provide information on sex, age, or reproductive class of the focal species.

Fig. 8.1. Analysis of stomach contents can reveal important information about food resources used by wildlife hazardous to aviation, such as these June beetles (*Phyllophaga* spp.) consumed by a laughing gull at an eastern U.S. airport. Photo credit: Brian E. Washburn

After samples are obtained, initial analysis usually consists of sorting and identifying all food items, and then summarizing the results in terms of frequency of occurrence, number, and volumetric proportions in the diet (Rosenberg and Cooper 1990), resulting in a ranked list of foods consumed. When identifying dietary samples, it is helpful to have a reasonably complete inventory of vegetation, small mammals, insects, or other potential food items present at the airport. Many airports have wildlife hazard management plans in place; these serve as good starting points for such inventories.

For many airport applications, it is likely not necessary to conduct detailed statistical analyses of food selection (i.e., quantifying food availability and comparing it to diet composition; see McDonald et al. 2005 for an overview of analysis methods). In this way, most studies of airport food habits are greatly simplified compared to many other investigations (Washburn et al. 2011). Even so, as noted above, one must consider food availability as it relates to diet composition

Table 8.1. Techniques for investigating wildlife food habits at airports. Note that examining the gastrointestinal tracts from animals killed during control activities or from birds struck by aircraft presents both advantages and disadvantages. Modified from Rosenberg and Cooper (1990), Litvaitis et al. (1994), and Litvaitis (2000).

Method	Advantages	Disadvantages
Gastrointestinal tracts	Can examine sex, age, physical condition, reproductive status, and other traits; samples can be readily obtained from animals killed during control activities or from animals struck by aircraft.	Samples are usually limited to animals that are killed during control activities or struck by aircraft; heavily masticated or partially digested materials can be difficult to identify.
Pellet or feces analysis	Inexpensive; makes it possible to sample a large proportion of the population; can be done with minimal disturbance; identification guides and keys of hair, mammal skulls, and the like, are available.	Usually cannot determine sex or age class of focal species; differential digestibility can bias relative importance of various foods; samples can be greatly fragmented.
Direct observation	Inexpensive; sex and age classes can sometimes be determined; birds are not disturbed; can sample a large proportion of the population.	Dense vegetation can obscure observations; biased toward large and conspicuous prey; quantity of food consumed can be difficult to estimate; if control is necessary, there is often limited time available to observe focal animals.
Feeding site surveys	Can identify major foods consumed by species of interest; can roughly estimate quantity of food consumed.	Completely consumed foods cannot be surveyed; usually cannot determine sex or age class of focal species; usually only applicable to herbivores.

Fig. 8.2. A ring-billed gull (*Larus delawarensis*) feeds on earthworms on an active aircraft taxiway. Direct observation of wildlife foraging in airport environments can identify important food resources. Photo credit: U.S. Department of Agriculture, Wildlife Services

before initiating management of food resources. Furthermore, a more robust analysis of food selection may be warranted when contemplating management (e.g., removal) of a valuable or sensitive resource, such as expensive landscaping plants established at the airport for aesthetic reasons.

In some circumstances, wildlife foraging can be observed directly, and the use of dietary samples can be bypassed entirely (Table 8.1; Fig. 8.2). Direct observation is inexpensive and minimally invasive, and at times sex and age classes can be determined. But data from direct observations are generally biased toward large and conspicuous prey (small prey items are often missed entirely), and observations are usually limited to open environments. In addition, if the focal individual presents an immediate risk of collision with an aircraft, dispersal or removal obviously takes precedence over observation. Apparent feeding sites also can be investigated to determine the species responsible for food consumption and the amount of food consumed. This technique is limited to certain circumstances (e.g., foraging on agricultural crops; MacGowan et al. 2006) and is usually applicable only to herbivores. When conducting direct observations or feeding site surveys, it is still important to consider food availability in relation to foods consumed.

Managing Wildlife Food Resources at or near Airports

Once the most important food resources used by hazardous wildlife at a given airport have been identified, they should be eliminated or modified if possible. In some cases this task might seem relatively straightforward, but other situations are more challenging. Every portion of the airport must be "covered" by some form of land use, and airport wildlife managers must ensure that the chosen replacement for a wildlife food resource does not present an even greater attractant or, worse yet, attract a different but more hazardous wildlife species. Even bare pavement can be an attractant for some birds (e.g., gulls; Belant et al. 1995).

The most effective management of wildlife food resources is conducted during the planning process, before land covers are established at airports (Washburn and Seamans 2004, Blackwell et al. 2009). Even considering the caveat on dietary breadth explored above, general food habit preferences are reasonably well known for most bird species that are hazardous to aircraft (the *Birds of North America* series is an excellent resource for North American species; http://bna.birds.cornell.edu/bna/). Many landscaping plants, turfgrasses, trees, and other potential food resources that are best avoided at airports can be eliminated from consideration before they are established. For example, most airports maintain relatively large expanses of turfgrass adjacent to taxiways and runways. The species composition, seed production capacity, and height of these turfgrass areas should be managed to minimize use of this resource by wildlife hazardous to aviation, especially Canada geese (DeVault et al. 2011). Fortunately, Canada geese have clear preferences for some turfgrass species over others (Conover 1991, Washburn et al. 2007, Washburn and Seamans 2012). Unpalatable turfgrass species should be given high consideration for establishment at airports (Chapter 10). Regardless, all airport planning and construction projects should be done in consultation with a knowledgeable airport wildlife biologist.

We list wildlife food resources commonly found at airports and give recommendations for management of those resources in Table 8.2 (see also FAA 2007). We also provide examples from the scientific literature that provide additional details regarding management of specific food resources. Several of the food resources listed (i.e., carrion, agricultural crops, and municipal solid waste) warrant further discussion, because they are particularly attractive to hazardous wildlife or are difficult to remove or manage appropriately.

Carrion (e.g., animals struck and killed by cars or aircraft) should be removed from airport grounds and disposed of immediately upon discovery (Blackwell and Wright 2006). Although vultures are the best-known scavengers (and are extremely hazardous to aircraft), nearly all carnivorous vertebrates will eat carrion (DeVault et al. 2003). Hawks, eagles, owls, crows, gulls, and carnivorous mammals are all attracted to animal carcasses, and all are unwanted at airports (Fig. 8.3, see p. 87).

Recent and ongoing research has suggested that some agricultural crops might be compatible with safe airport operations (see below; Chapter 11). However, other crops like corn (*Zea mays*) and small grains like wheat (*Triticum* spp.) are known wildlife attractants (Cerkal et al. 2009) and should be avoided at and near airports when possible. Many wildlife species attracted to corn and small grains are especially hazardous to aircraft, such as Canada geese, snow geese (*Anser caerulescens*), and sandhill cranes (*Grus canadensis*), and large flocks of blackbirds (e.g., red-winged blackbirds [*Agelaius phoeniceus*]; DeVault et al. 2011). Unfortunately, cultivation of corn and small grains is surprisingly common at airports (especially smaller facilities; DeVault et al. 2009). Further information on crop production that is safe for airport use is needed (Chapter 11).

Municipal solid waste management facilities, such as open landfills and trash-transfer stations, can attract birds hazardous to aviation and can increase the potential for strikes when these facilities are located near airports (Fig. 8.4, see p. 87). Gulls, vultures, European starlings, rock pigeons, and other birds forage on anthropogenic food waste at landfills and trash-transfer facilities (Patton 1988, Washburn 2012). Guidance and regulations regarding the siting of waste management facilities related to airports are available (FAA 2000, 2007). Considerable variation exists among solid waste management facilities with regard to their attractiveness to hazardous wildlife. Foraging birds heavily use some facilities, whereas wildlife use of other facilities is essentially nonexistent. Washburn (2012) found that several factors, including the geographic location, time of year, building design, and on-site facility management practices (e.g., cleanliness of outside areas), interact to influence the attractiveness of trash-transfer stations to hazardous wildlife. Integrated wildlife damage management practices that involve active wildlife control

Table 8.2. Common wildlife food resources found at and near airports, the hazardous species they attract, and options for management.

Food resource	Species or species group	Management options	References
Turfgrasses	Canada geese	Replace palatable turfgrasses with less desired species or types, alternative land covers, or artificial turf.	Conover (1991), Pochop et al. (1999), Washburn et al. (2007), Washburn and Seamans (2012); Chapter 10
Other terrestrial vegetation (seeds, fruit, etc.)	White-tailed deer, passerine birds, doves and pigeons, wild turkeys	Remove plants; erect netting or fencing.	Bernhardt et al. (2009), Biondi et al. (2011)
Aquatic vegetation	Ducks	Remove plants; erect netting; physically alter stormwater retention and detention ponds.	Blackwell et al. (2008); Chapter 9
Small grain and corn production	Geese, blackbirds, doves, sandhill cranes	Convert to alternative crops.	Williams and Jackson (1981), Humberg et al. (2007), Blackwell et al. (2009), Martin et al. (2011); Chapter 11
Small mammals	Raptors, owls, coyotes	Reduce population with rodenticides; manage or convert vegetation.	Stucker and Dunlap (2002), Witmer and Fantinato (2003), Witmer et al. (2007), Witmer (2011)
Carrion	Nearly all carnivorous vertebrates, but especially vultures, gulls, raptors, crows, coyotes, raccoons	Promptly remove and dispose of vertebrates struck by aircraft or ground vehicles.	DeVault et al. (2003), Blackwell and Wright (2006)
Fish and other aquatic animals	Ducks, osprey, eagles, pelicans, cormorants, herons	Remove fish in airport water bodies.	Werner and Dorr (2006)
Earthworms	Gulls, passerine birds	Modify runways and taxiways; use earthworm deterrents.	Dekker (2003)
Insects	Gulls, passerine birds, some raptors	Modify vegetation.	Buckley and McCarthy (1994), Caccamise et al. (1994), Bernhardt et al. (2010), Kutschbach-Brohl et al. (2010), Washburn et al. (2011)
Trash facilities (landfills, trash-transfer stations)	European starlings, gulls, pigeons	Properly manage trash facilities; employ frightening (dispersal) techniques.	Belant (1997), Patton (1988), Washburn (2012)
Human food waste (restaurants, etc.)	Geese, ducks, European starlings, pigeons, sparrows, and so on	Discourage feeding wildlife near airports.	International Civil Aviation Organization (2002)

(e.g., hazing with pyrotechnics) and alteration of facility operations (e.g., covering waste at night) can be effective in reducing the use of waste management facilities by hazardous birds.

Research Needs

The management of wildlife food resources at airports is inextricably linked to management of habitats (Caccamise et al. 1994, Barras and Seamans 2002, Kutschbach-Brohl et al. 2010, Washburn et al. 2011, Witmer 2011) and, unfortunately, the practice of habitat management at airports is often based on longstanding paradigms with scant scientific support (Blackwell et al. 2013; Chapter 11). For example, the FAA categorically denounces the presence of agriculture (including hay crops) on airport properties in the USA, because many types of agriculture provide food resources and thus attract hazardous wildlife species (FAA 2007; see also similar recommendations by the International Civil Aviation Organization 2002). However, field studies examining the importance of various types of agriculture in the diets of hazardous wildlife are lacking (Blackwell et al. 2009, Martin

Fig. 8.3. A peregrine falcon (*Falco peregrinus*) feeds on a gull carcass at a major airport in the eastern USA. Animal carcasses found on airfields should be removed immediately upon discovery so they do not attract hazardous wildlife. Photo credit: Jenny Mastantuono

Fig. 8.4. When not managed properly, solid waste management facilities like this trash-transfer station can attract wildlife hazardous to aviation. Photo credit: Brian E. Washburn

et al. 2011). Re-examination of habitat management paradigms (and wildlife food availability) will be needed to advance the science of wildlife management at airports. Given the immense scale of managed land at airports worldwide (e.g., airport grasslands in the continental USA encompass >3,300 km² [1,274 miles²], and the USA contains about 15,000 of the world's 44,000 airports; U.S. Central Intelligence Agency 2010, DeVault et al. 2012), questions regarding the interplay among habitat types, food availability, and wildlife movements on and near airports are critical.

Summary

Food acquisition is often the chief motivation for wildlife to use airport habitats. Frequently, the most effective way to reduce populations of hazardous wildlife at airports is to determine which foods are being used and then remove or modify those foods to make them less attractive. Several techniques are available for determining food selection by wildlife at airports, and samples are often readily available (e.g., animals struck by aircraft or collected during control activities). Once important food resources have been identified, management actions can be employed to reduce or remove them. Given the variety and abundance of wildlife foods available at airports, such efforts can be difficult and require careful consideration of the proper management actions to implement, such as habitat manipulation. Integrated wildlife damage management practices can be effective in removing food attractants from airport environments and reduce the risk of damaging wildlife strikes.

LITERATURE CITED

Baker, J. A., and R. J. Brooks. 1981a. Distribution patterns of raptors in relation to density of meadow voles. Condor 83:42–47.

Baker, J. A., and R. J. Brooks. 1981b. Raptor and vole populations at an airport. Journal of Wildlife Management 45:390–396.

Barras, S. C., and T. W. Seamans. 2002. Habitat management approaches for reducing wildlife use of airfields. Proceedings of the Vertebrate Pest Conference 20:309–315.

Bechard, M. J. 1982. Effect of vegetative cover on foraging site selection by Swainson's hawk. Condor 84:153–159.

Beier, P., and J. E. Drennan. 1997. Forest structure and prey abundance in foraging areas of northern goshawks. Ecological Applications 7:564–571.

Belant, J. L. 1997. Gulls in urban environments: landscape-level management to reduce conflict. Landscape and Urban Planning 38:245–258.

Belant, J. L., S. W. Gabrey, R. A. Dolbeer, and T. W. Seamans. 1995. Methyl anthranilate formulations repel gulls and mallards from water. Crop Protection 14:171–175.

Bernhardt, G. E., L. Kutschbach-Brohl, B. E. Washburn, R. B. Chipman, and L. C. Francoeur. 2010. Temporal variation in terrestrial invertebrate consumption by laughing gulls in New York. American Midland Naturalist 163:442–454.

Bernhardt, G. E., Z. J. Patton, L. A. Kutschbach-Brohl, and R. A. Dolbeer. 2009. Management of bayberry in relation to tree-swallow strikes at John F. Kennedy International Airport, New York. Human–Wildlife Conflicts 3:237–241.

Biondi, K. M., J. L. Belant, J. A. Martin, T. L. DeVault, and G. Wang. 2011. White-tailed deer incidents with U.S. civil aircraft. Wildlife Society Bulletin 35:303–309.

Blackwell, B. F., T. L. DeVault, E. Fernández-Juricic, and R. A. Dolbeer. 2009. Wildlife collisions with aircraft: a missing component of land-use planning for airports. Landscape and Urban Planning 93:1–9.

Blackwell, B. F., L. M. Schafer, D. A. Helon, and M. A. Linnell. 2008. Bird use of stormwater-management ponds: decreasing avian attractants on airports. Landscape and Urban Planning 86:162–170.

Blackwell, B. F., T. W. Seamans, P. M. Schmidt, T. L. DeVault, J. L. Belant, M. J. Whittingham, J. A. Martin, and E. Fernández-Juricic. 2013. A framework for managing airport grasslands and birds amidst conflicting priorities. Ibis 155:199–203.

Blackwell, B. F., and S. E. Wright. 2006. Collisions of red-tailed hawks (Buteo jamaicensis), turkey (Cathartes aura), and black vultures (Coragyps atratus) with aircraft: implications for bird strike reduction. Journal of Raptor Research 40:76–80.

Buckley, N. J. 1999. Black vulture (Coragyps atratus). Number 411 in A. Poole and F. Gill, editors. The birds of North America. Academy of Natural Sciences, Philadelphia, Pennsylvania, and American Ornithologists' Union, Washington, D.C., USA.

Buckley, P. A., and M. G. McCarthy. 1994. Insects, vegetation, and the control of laughing gulls (Larus atricilla) at Kennedy International Airport, New York City. Journal of Applied Ecology 31:291–302.

Caccamise, D. F., J. J. Dosch, K. Benett, L. M. Reed, and L. DeLay. 1994. Management of bird strike hazards at airports: a habitat approach. Proceedings of the International Bird Strike Committee 22:285–306.

Cerkal, R., K. Vejražka, J. Kamler, and J. Dvořák. 2009. Game browse and its impact on selected grain crops. Plant, Soil, and Environment 55:181–186.

Coleman, J. S., and J. D. Fraser. 1987. Food habits of black and turkey vultures in Pennsylvania and Maryland. Journal of Wildlife Management 51:733–739.

Coleman, J. S., and J. D. Fraser. 1989. Habitat use and home ranges of black and turkey vultures. Journal of Wildlife Management 53:782–792.

Conover, M. R. 1991. Herbivory by Canada geese: diet selection and effects on lawns. Ecological Applications 1:231–236.

Dekker, A. 2003. Taking habitat management one step further. Proceedings of the International Bird Strike Committee 26:265–272.

DeVault, T. L., J. L. Belant, B. F. Blackwell, J. A. Martin, J. A. Schmidt, L. W. Burger Jr., and J. W. Patterson Jr. 2012. Airports offer unrealized potential for alternative energy production. Environmental Management 49:517–522.

DeVault, T. L., J. L. Belant, B. F. Blackwell, and T. W. Seamans. 2011. Interspecific variation in wildlife hazards to aircraft: implications for airport wildlife management. Wildlife Society Bulletin 35:394–402.

DeVault, T. L., and A. R. Krochmal. 2002. Scavenging by snakes: an examination of the literature. Herpetologica 58:429–436.

DeVault, T. L., J. E. Kubel, O. E. Rhodes Jr., and R. A. Dolbeer. 2009. Habitat and bird communities at small airports in the midwestern USA. Proceedings of the Wildlife Damage Management Conference 13:137–145.

DeVault, T. L., B. D. Reinhart, I. L. Brisbin Jr., and O. E. Rhodes Jr. 2004. Home ranges of sympatric black and turkey vultures in South Carolina. Condor 106:706–711.

DeVault, T. L., B. D. Reinhart, I. L. Brisbin Jr., and O. E. Rhodes Jr. 2005. Flight behavior of black and turkey vultures: implications for reducing bird–aircraft collisions. Journal of Wildlife Management 69:601–608.

DeVault, T. L., O. E. Rhodes Jr., and J. A. Shivik. 2003. Scavenging by vertebrates: behavioral, ecological, and evolutionary perspectives on an important energy transfer pathway in terrestrial ecosystems. Oikos 102:225–234.

Dolbeer, R. A., and S. E. Wright. 2009. Safety management systems: how useful will the FAA National Wildlife Strike Database be? Human–Wildlife Conflicts 3:167–178.

Dolbeer, R. A., S. E. Wright, and E. C. Cleary. 2000. Ranking the hazard level of wildlife species to aviation. Wildlife Society Bulletin 28:372–378.

Dolbeer, R. A., S. E. Wright, J. Weller, and M. J. Begier. 2010. Wildlife strikes to civil aircraft in the United States 1990–2009. Serial Report 16. U.S. Department of Transportation, Federal Aviation Administration, Office of Airport Safety and Standards, Washington, D.C., USA.

FAA. Federal Aviation Administration. 2000. Construction or establishment of landfills near public airports. Advisory Circular 150/5200-34. U.S. Department of Transportation, Washington, D.C., USA.

FAA. Federal Aviation Administration. 2007. Hazardous wildlife attractants on or near airports. Advisory Circular 150/5200-33B. U.S. Department of Transportation, Washington, D.C., USA.

Fischl, J., and D. F. Caccamise. 1987. Relationships of diet and roosting behavior in the European starling. American Midland Naturalist 117:395–404.

Grant, T. A., E. M. Madden, T. L. Shaffer, P. J. Pietz, G. B. Berkey, and N. J. Kadrmas. 2006. Nest survival of clay-colored and vesper sparrows in relation to woodland edge in mixed-grass prairies. Journal of Wildlife Management 70:691–701.

Heinrich, G., and K. R. Studenroth Jr. 1996. Agkistrodon piscivorus conanti (Florida cottonmouth): diet. Herpetological Review 27:22.

Humberg, L. A., T. L. DeVault, B. J. MacGowan, J. C. Beasley, and O. E. Rhodes Jr. 2007. Crop depredation by wildlife in northcentral Indiana. Proceedings of the National Wild Turkey Symposium 9:59–65.

International Civil Aviation Organization. 2002. Land use and environmental control. Airport Planning Manual, Document 9184 AN/902, Part 2. Montreal, Quebec, Canada.

Johnson, D. H. 1980. The comparison of usage and availability measurements for evaluating resource preference. Ecology 61:65–71.

Kelly, N. E., D. W. Sparks, T. L. DeVault, and O. E. Rhodes Jr. 2007. Diet of black and turkey vultures in a forested landscape. Wilson Journal of Ornithology 119:267–270.

Kirk, D. A., and M. J. Mossman. 1998. Turkey vulture (Cathartes aura). Number 339 in A. Poole and F. Gill, editors. The birds of North America. Academy of Natural Sciences, Philadelphia, Pennsylvania, and American Ornithologists' Union, Washington, D.C., USA.

Krapu, G. L., D. A. Brandt, and R. R. Cox Jr. 2004. Less waste corn, more land in soybeans, and the switch to genetically modified crops: trends with important implications for wildlife management. Wildlife Society Bulletin 32:127–136.

Kutschbach-Brohl, L., B. E. Washburn, G. E. Bernhardt, R. B. Chipman, and L. C. Francoeur. 2010. Arthropods of a semi-natural grassland in an urban environment: the John F. Kennedy International Airport, New York. Journal of Insect Conservation 14:347–358.

Lima, S. L. 1998. Non-lethal effects in the ecology of predator–prey interactions. BioScience 48:25–34.

Lima, S. L., and L. M. Dill. 1990. Behavioural decisions made under the risk of predation: a review and prospectus. Canadian Journal of Zoology 68:619–640.

Litvaitis, J. A. 2000. Investigating food habits of terrestrial vertebrates. Pages 165–190 in L. Boitani and T. K. Fuller, editors. Research techniques in animal ecology. Columbia University Press, New York, New York, USA.

Litvaitis, J. A., K. Titus, and E. Anderson. 1994. Measuring vertebrate use of terrestrial habitats and food. Pages 254–274 in T. Bookhout, editor. Research and management techniques for wildlife and habitats. The Wildlife Society, Bethesda, Maryland, USA.

MacArthur, R. H., and E. R. Pianka. 1966. On the optimal use of a patchy environment. American Naturalist 100:603–609.

MacGowan, B. J., L. A. Humberg, J. C. Beasley, and O. E. Rhodes Jr. 2006. Identification of wildlife crop damage. Publication FNR-267-W. Purdue University Extension, West Lafayette, Indiana, USA.

Martin, J. A., J. L. Belant, T. L. DeVault, L. W. Burger Jr., B. F. Blackwell, S. K. Riffell, and G. Wang. 2011. Wildlife risk to aviation: a multi-scale issue requires a multi-scale solution. Human–Wildlife Interactions 5:198–203.

McDonald, L. L., J. R. Alldredge, M. S. Boyce, and W. P. Erickson. 2005. Measuring availability and vertebrate use of terrestrial habitats and foods. Pages 465–488 in C. E. Braun, editor. Techniques for wildlife investigations and management. The Wildlife Society, Bethesda, Maryland, USA.

Parker, K., and M. L. McCallum. 2010. *Pantherophis obsoletus* (Texas ratsnake): diet and feeding behavior. Herpetological Review 41:502.

Patton, S. R. 1988. Abundance of gulls at Tampa Bay landfills. Wilson Bulletin 100:431–442.

Pietz, P. J., and D. A. Granfors. 2000. Identifying predators and fates of grassland passerine nests using miniature video cameras. Journal of Wildlife Management 64:71–87.

Pochop, P. A., J. L. Cummings, K. L. Wedemeyer, R. M. Engeman, and J. E. Davis Jr. 1999. Vegetation preferences of captive Canada geese at Elmendorf Air Force Base, Alaska. Wildlife Society Bulletin 27:734–740.

Polis, G. A. 1991. Complex trophic interactions in deserts: an empirical critique of food-web theory. American Naturalist 138:123–155.

Preston, C. R. 1990. Distribution of raptor foraging in relation to prey biomass and habitat structure. Condor 92:107–112.

Rolando, A. 2002. On the ecology of home range in birds. Review d'Ecologie—La Terre et la Vie 57:53–73.

Rosenberg, K. V., and R. J. Cooper. 1990. Approaches to avian diet analysis. Studies in Avian Biology 13:80–90.

Savidge, J. A. 1988. Food habits of *Boiga irregularis*, an introduced predator on Guam. Journal of Herpetology 22:275–282.

Shoener, T. 1971. Theory of feeding strategies. Annual Review of Ecology and Systematics 2:369–404.

Shoener, T. 1986. A brief history of optimal foraging theory. Pages 5–68 in A. C. Kamil, J. R. Krebs, and H. R. Pulliam, editors. Foraging behavior. Plenum, New York, New York, USA.

Sieving, K. E., and M. F. Willson. 1998. Nest predation and avian species diversity in northwestern forest understory. Ecology 79:2391–2402.

Sterner, R. T., D. J. Elias, M. V. Garrison, B. E. Johns, and S. R. Kilburn. 1984. Birds and airport agriculture in the conterminous United States: a review of literature. Pages 319–329 in M. J. Harrison, S. A. Gauthreaux Jr., and L. A. Abron-Robinson, editors. Proceedings of the wildlife hazards to aircraft conference. DOT/FAA/AAS/84-1. U.S. Department of Transportation, Federal Aviation Administration, Office of Airport Standards. Washington, D.C., USA.

Stucker, K. P., and B. G. Dunlap. 2002. Food habits of raptors using airports in north-central Kentucky. Proceedings of the Vertebrate Pest Conference 20:170–174.

U.S. Central Intelligence Agency. 2010. The world factbook. https://www.cia.gov/library/publications/the-world-factbook/geos/us.html.

Wakely, J. S. 1978. Factors affecting the use of hunting sites by ferruginous hawks. Condor 80:316–326.

Washburn, B. E. 2012. Avian use of solid waste transfer stations. Landscape and Urban Planning 104:388–394.

Washburn, B. E., S. C. Barras, and T. W. Seamans. 2007. Foraging preferences of captive Canada geese related to turfgrass mixtures. Human–Wildlife Conflicts 1:214–223.

Washburn, B. E., G. E. Bernhardt, and L. A. Kutschbach-Brohl. 2011. Using dietary analyses to reduce the risk of wildlife–aircraft collisions. Human–Wildlife Interactions 5:204–209.

Washburn, B. E., and T. W. Seamans. 2004. Management of vegetation to reduce wildlife hazards at airports. Proceedings of the 2004 worldwide airport technology transfer conference. Federal Aviation Administration, April 2004, Atlantic City, New Jersey, USA.

Washburn, B. E., and T. W. Seamans. 2012. Foraging preferences of Canada geese among turfgrasses: implications for reducing human–goose conflicts. Journal of Wildlife Management 76:600–607.

Werner, S. J., and B. S. Dorr. 2006. Influence of fish stocking density on the foraging behavior of double-crested cormorants, *Phalacrocorax auritus*. Journal of the World Aquaculture Society 37:121–125.

Williams, G. E., and P. B. Wood. 2002. Are traditional methods of determining nest predators and nest fates reliable? An experiment with wood thrushes (*Hylocichla mustelina*) using miniature video cameras. Auk 119:1126–1132.

Williams, R. E., and W. B. Jackson. 1981. Dietary comparisons of red-winged blackbirds, brown-headed cowbirds, and European starlings in north-central Ohio. Ohio Journal of Science 81:217–225.

Witmer, G. W. 2011. Rodent population management at Kansas City International Airport. Human–Wildlife Interactions 5:269–275.

Witmer, G. W., and J. W. Fantinato. 2003. Management of rodent populations at airports. Proceedings of the Wildlife Damage Management Conference 10:350–357.

Witmer, G. W., R. Saylor, D. Huggins, and J. Capelli. 2007. Ecology and management of rodents in no-till agriculture in Washington, USA. Integrative Zoology 2:154–164.

9

BRADLEY F. BLACKWELL
DAVID FELSTUL
THOMAS W. SEAMANS

Managing Airport Stormwater to Reduce Attraction to Wildlife

An airport is a component of the landscape, contributing to and subject to local- and landscape-level factors that affect wildlife populations and the hazards that these species pose to aviation (Blackwell et al. 2009, Martin et al. 2011). Water resources at and near an airport, in the form of both surface water and contained runoff, are recognized by the Federal Aviation Administration (FAA) as potential attractants to wildlife that pose hazards to aviation safety (FAA 2007). Surface water, including aboveground stormwater detention/retention facilities (see U.S. Environmental Protection Agency 2006), can represent a substantial proportion of the area within siting criteria for U.S. airports. An analysis of water coverage at 49 certificated airports (FAA 2004) revealed that surface water composed on average 6.0% (standard deviation [SD] = 10.4%, range = 0.04–48.3%; B. F. Blackwell, unpublished data) of the area within the 3-km [1.9-mile] FAA siting criteria ($\bar{X}$ = 275 ha, SD = 511 ha). A recent analysis of bird–aircraft strike data for avian species involved in at least 50 total strikes reported to the FAA (1990–2008; summarized in FAA 2011) revealed that 13 of the 52 species (25%) have foraging and breeding ecologies primarily associated with water (Blackwell et al. 2013). Moreover, these 13 species were responsible for >51% of damaging strikes (Dolbeer et al. 2000, DeVault et al. 2011) during this period.

Given the obvious necessity of water as a resource to wildlife and the relative aviation hazards posed by bird species whose life histories are tied to water, aspects of species ecology should inform airport biologists in the management of natural or constructed water resources to reduce attractive features. Likewise, informed exchange between airport biologists and engineers responsible for the design of runoff containment and treatment facilities will yield facilities that minimize attractant features to birds. Our purpose for this chapter is to demonstrate how airport stormwater runoff can be managed effectively to reduce or prevent the establishment of a resource on and near airport properties. We discuss features of water resources that attract birds, describe common operational conditions at airports with regard to managing stormwater runoff, and review findings on postconstruction methods to deter bird use of stormwater facilities. In addition, we review advantages and disadvantages of novel runoff containment systems for airfields, as well as considerations for stormwater management outside of the air operations area (AOA) but within or proximate to FAA siting criteria.

Birds and Water

Short of thirst, no single factor drives avian use of water resources. Commonalities observed in avian use of natural and constructed systems, however, are important to how airport authorities plan for and manage their water resources to reduce use by birds. Within wetland systems, avian species richness is positively correlated with wetland complexes (20–30 ha for marsh and >55 ha of marsh complex within 5 km [3 miles]), as opposed to larger (up to 180 ha), isolated marshes (Brown and Dinsmore 1986; see also Fairbairn and Dinsmore 2001). Also, wetlands with an intermediate level of emergent cover (33–66%) have been found to harbor greater species richness (Belánger and Couture 1988, Gibbs et al. 1991, Creighton et al. 1997). Working with lake systems, Suter (1994) linked abundance and richness of various avifauna populations to area, food availability, and shoreline vegetation complexity. In addition, overall mean and maximum species richness increased with nutrient load, as did maximum bird densities among guilds. Similar conditions are possible within stormwater impoundments (ponds and reservoirs) with sediment deposits accumulating over time, resulting in vegetation complexes that can support an array of invertebrate and vertebrate diversity (Le Viol et al. 2009).

In a broad sense, bird use of water resources is driven primarily by site-specific relationships of system, area, cover, food resources, and complexity with regard to neighboring resources. Recent findings for bird use of stormwater management ponds are similar to those for natural systems. Modeling avian use of stormwater management ponds in the Pacific Northwest region of the USA, which served as surrogates to those at airport facilities, revealed that surface area available for water containment, area of open water available, pond perimeter, and pond isolation were factors that predicted use by nine of 13 considered bird groups (within Accipitridae, Anatidae, Ardeidae, Charadriidae, Columbidae, Laridae, and Rallidae; Blackwell et al. 2008). Posthoc modeling by the authors revealed that the probability of pond use by birds considered hazardous to aviation (Dolbeer et al. 2000, DeVault et al. 2011) was about 100% when perimeter irregularity (i.e., the quotient associated with the ratio of pond perimeter to perimeter of a perfect circle of equal area) equaled 7. In contrast, the probability of use by birds hazardous to aviation was near zero when the facility was isolated (>8 km [5 miles] horizontal distance) from other surface-water resources.

In effectively incorporating the information discussed above with guidance on airport stormwater management, one must first understand that stormwater runoff poses multiple safety and regulatory challenges for airport managers.

Stormwater Management Practices at Airports

At U.S. airports, the immediate safety of maneuvering aircraft and water quality are the predominant concerns of FAA guidance for runoff management. Regulatory control of water-quality practices at airports stem from National Pollution Discharge Elimination System requirements under the U.S. Clean Water Act and local ordinances (FAA 2006). Best management practices (BMPs) associated with stormwater containment consider site-specific physical conditions, area of watershed (including area of impermeable surfaces on and near airport property), runoff volume or peak flow, and water-quality objectives (FAA 2006, Goff and Gentry 2006). BMP designs that can attract wildlife, particularly birds, generally require some period of exposed storage or "ponding" of runoff. These designs at airports include extended dry detention ponds intended to store runoff after a storm event for up to 48 hr; retention ponds that serve dual purposes of containing water from a storm event and treating the runoff for pollutant removal; and infiltration basins in which stored water is exfiltrated through permeable soils (FAA 2006). In addition, FAA (2008) recommends conversion of "suitable unused airport land" to lagoons and retention ponds to facilitate the collection of large volumes of glycol-based fluid waste (i.e., deicing chemicals); in this case the potential creation of a wildlife resource is not considered. However, using ponds to contain deicing chemicals poses disadvantages, in addition to possibly attracting wildlife, that are associated with effective product recovery or treatment (see Airport Cooperative Research Program 2009).

For any exposed containment of stormwater runoff, airport managers are directed to FAA (2007) for guidance on wildlife hazards, where suggested techniques focus on reducing wildlife (primarily bird) access via use of synthetic covers, floating covers, netting, or wire grids (see also International Civil Aviation Organization 1991:11–12). But these postconstruction techniques can be costly with regard to purchase, installation, and maintenance, and efficacy is not always clear. For example, overhead wires or lines in various arrangements have been effective in repelling a variety of birds (McAtee and Piper 1936; Amling 1980; Blokpoel and Tessier 1983, 1984; Forsythe and Austin 1984;

Fig. 9.2. Netted reservoir near Seattle, Washington, USA. The resource is protected physically, but it still serves as a visual attractant. Photo credit: Mike Linnell

Fig. 9.1. Mallards (*Anas platyrhynchos*) under wires. Photo credit: Greg Martinelli

McLaren et al. 1984; Dolbeer et al. 1988; Pochop et al. 1990), but efficacy is site specific. Pochop et al. (1990) noted that bird reaction to overhead lines varies by species, spacing, attractiveness of sites protected, age of birds, and possibly height of lines above the protected area (Fig. 9.1).

Anthony Duffiney (U.S. Department of Agriculture, Wildlife Services, unpublished data) found that the number of mute swans (*Cygnus olor*), gulls (Laridae), Canada geese (*Branta canadensis*), and most waterfowl species using containment ponds (~15.4 ha) at Detroit Metro Airport, Romulus, Michigan, USA, was reduced after installation of parallel steel wires at 30.5-m (100-foot) intervals, supported by metal posts on shore. However, icing and increased tension on the wires, as well as damage to supports during mowing, necessitated frequent year-round maintenance. In another airport application, a wire grid system installed to deter ducks, primarily mallards (*Anas platyrhynchos*), from drainages proved too costly with regard to equipment and maintenance, and effective control over all points of entry was not achieved (A. Baxter, U.K. Food and Environment Research Agency, unpublished data). When a 15-m (50-foot) grid system was installed over 2-ha wastewater ponds in North Carolina, USA, the total number of waterfowl using the ponds surprisingly increased. Canada goose numbers declined, while mallard, ring-necked duck (*Aythya collaris*), and ruddy duck (*Oxyura jamaicensis*) numbers (among other species) increased (T. W. Seamans, U.S. Department of Agriculture, unpublished data). In this case, enhanced protection from avian predators—or added resource

value due to aggregations of conspecifics (e.g., Arengo and Baldassarre 2002) and absence of larger, dominant, or competitive species—could have contributed to the attractiveness of the site.

Completely covering exposed water containment systems physically and visually (e.g., via synthetic cover or floating devices that cover the exposed pond surface area) is likely the only means of effectively reducing the attraction to birds (Fig. 9.2). However, cover alternatives pose problems, as well. To our knowledge there is no candidate vegetation that might minimize available surface area of water to birds, survive both flooding and water drawdown, and not provide food, roosting, or nesting resources. Complete coverage of standing water via synthetic or floating covers will reduce solar radiation, an important factor in the control of bacterial growth (Davies and Bavor 2000), and can negatively affect pond hydraulics, oxygenation, and biological activity (e.g., see effects of pond ice cover; Semadeni-Davies 2006). Water quality in natural receiving systems might subsequently be degraded.

Management of stormwater runoff at airports to enhance aircraft safety, to achieve water-quality goals, and to minimize attractants to birds and other wildlife is complex, if not contradictory, begging the question as to whether alternatives exist that meet BMP requirements for controlling airport stormwater runoff.

Potential Alternatives

Higgins and Liner (2007) noted that containment and treatment of stormwater at airports, particularly runoff contaminated with deicing chemicals, via conventional means (e.g., ponds) is particularly difficult when conditions are cold and runoff is intermittent and at high volumes over short periods. However, the authors reported an "innovative approach" using aerated gravel beds known as subsurface flow wetlands (SSFWs). According to the authors, SSFWs are insulated, aerated, easy to operate, and their construction, operation, and maintenance costs are a fraction of those at alternative conventional stormwater treatment facilities (< 50%). As to wildlife hazards, SSFWs are underground and thus do not attract avian species. The authors note installations only at Edmonton International Airport (Edmonton, Alberta, Canada), Heathrow International Airport (London, United Kingdom), and Air Express Airport (Wilmington, Ohio, USA). The first two installations are horizontal flow SSFWs, while the third is a reciprocating flow (tidal), vertical flow SSFW. All three are associated with surge ponds in front of their multiple wetland basins (cells). Higgins and Liner (2007) also recognized problems associated with constructed wetlands, particularly those intended to treat glycol-contaminated stormwater runoff, as the wetlands tend to be large. A horizontal flow SSFW, like that at Heathrow, can experience plugging problems (e.g., due to freezing) in the shallow gravel of the primary cells.

As an alternative, Higgins and Liner (2007) recommended engineered wetlands known as semipassive constructed wetlands, designed so that operating and process conditions can be controlled, in contrast to the more passive operation of traditionally constructed wetlands. They suggest that engineered wetlands will allow higher levels of contaminant removals at higher throughputs and with much shorter residence times. The authors point to Buffalo Niagara International Airport, Buffalo, New York, USA, and its use of an aerated, vertical flow SSFW, engineered wetland, in which blowers introduce air under a gravel substrate 1.2–3.6 m (4–12 feet) thick. Aeration is directed upward through the gravel from a buried, fine-bubble diffusion system, countercurrent to downward percolating wastewater. The vegetated gravel surfaces of engineered wetlands are also insulated with layers of mulch or compost to prevent freezing, and the systems are designed to operate throughout northern winters and associated ambient air temperatures. In a controlled greenhouse experiment comparing performance of "surface flow, constructed wetlands" versus "subsurface flow, constructed wetlands" (essentially a SSFW, as described above) for treatment of synthetic sewer overflows, nitrogen, phosphorous, and chemical oxygen demand were removed faster by SSFWs and, in general, the end concentrations of the investigated pollutants were lower than in the surface flow constructed wetlands (Van de Moortel et al. 2009).

However, runoff management via SSFWs, or even belowground vaults for water containment, will not suffice for all locations. Other promising alternatives to control stormwater runoff that will satisfy both stormwater permit requirements and allow for safe operations at airports are a family of BMPs known collectively as low-impact development (LID; Dietz 2007, Davis 2008, Dietz and Clausen 2008) or green infrastructure (GI; see also Washington State Department of Transportation 2009; Oregon Department of Environmental Quality 2011a). The language of stormwater permits (U.S. Environmental Protection Agency 2012) defines these two approaches. Specifically, LID promotes the use of natural systems for infiltration, evapotranspiration, and reuse of rainwater, and can occur at a wide range of landscape scales (i.e., regional, community, and site). Similarly, GI is a comprehensive approach to water-quality protection defined by a range of natural and built systems and practices that use or mimic natural hydrologic processes to infiltrate, evapotranspirate, or reuse stormwater runoff at the site where the runoff is generated.

U.S. Environmental Protection Agency (2012) has organized LID/GI techniques into a number of categories, some of which are less applicable than others to airports, although all have techniques that are useful. Below we provide descriptions of the types of facility in each category and some general advantages and disadvantages to their use at airports. Every airport site is unique, however, and should be fully investigated before locating an LID stormwater facility on the airport property.

Category 1, Conservation Designs, includes measures such as preserving open space, clustering development, and using "skinny" streets. For airports, operational concerns largely determine layout and pavement extent. However, clustering stormwater facilities on one side and away from the runway (as per FAA 2012; see also FAA 2006) might be one type of conservation design appropriate at an airport. Clustering should decrease the frequency of wildlife crossing operational space. In addition, stormwater facilities should be located on the same side as natural attractants, such as wetlands, rivers, roosting trees, and food sources. As a caveat, we note recommendations by Blackwell et al. (2008) relative to minimizing density of water resources in locating stormwater management ponds.

Category 2, Infiltration Practices, includes infiltration trenches, porous pavement, and rain gardens— methods that depend upon relatively quick and efficient drainage. Infiltration trenches are long, narrow, stone-filled trenches used for the collection, temporary storage, and infiltration of stormwater runoff to groundwater. Standard infiltration trench designs work well in airport environments. Depending on the trench dimensions, the facility might be considered an underground injection control device (i.e., any subsurface drain fields that release fluids underground), subject to additional permitting requirements (see also U.S. Environmental Protection Agency 1999, 2003).

Porous pavement is an open-graded concrete or asphalt mix placed in a manner that results in a high degree of interstitial spaces or voids within the cemented aggregate. This technique demonstrates a high volume of absorption or storage within the voids and infiltration to subsoils. The pavement might be permeable concrete or asphalt, manufactured systems such as interlocking brick, or a combination of sand and brick lattice. At airports, porous pavement is suitable for passenger parking areas or service roads that are used occasionally. Concerns about weight-bearing capacity (FAA 2009) generally will not allow its use where aircraft are maneuvering or parking, including runway, taxiway, and clearway. In colder climates, the use of porous pavement in areas where grit is applied for traction, such as on parking lots, can result in pore clogging, standing water, or icy conditions.

Another infiltration practice, the rain garden, is an excavated depression, usually back-filled with an amended soil mixture and planted with a variety of native plants that tolerate saturated soils. Most rain gardens are constructed with up to 0.3 m (1 foot) of freeboard above the soil surface, which provides temporary ponding until runoff can infiltrate. A selling point of rain gardens emphasizes their wildlife habitat benefits from the plantings (food, shelter, nesting space). Coupled with the potential for extended ponding, however, rain gardens can become undesired wildlife attractants.

Minus the "garden" plantings, the facility would function similarly to an infiltration basin, providing the desired infiltration with a lower risk of attracting wildlife.

Key considerations for Infiltration Practices include siting where soils provide good infiltration during wet weather and adequate maintenance to prevent clogging. Infiltration facilities should not be used in areas with high groundwater tables, which might be the case for airport facilities located next to water bodies. These techniques also require extra pretreatment to remove solids that might clog the facility and cause ponding.

Category 3, Runoff Storage Practices, includes the use of rain barrels, cisterns, and green roofs, and works best in areas that can have substantial rainfall during warmer, typically drier months, such as the midwestern and southeastern USA. A rain barrel can capture runoff from a thunderstorm and be used for irrigation within days or weeks. Airports irrigate vegetation around terminals, and these types of storage methods can be connected to irrigation systems, lowering labor requirements while containing runoff that might pool elsewhere or be conveyed to stormwater management ponds on site. In climates such as the Pacific Northwest, the majority of rainfall occurs in winter, when soils are saturated and many plants are dormant. Capturing and storing most of the rainfall from the winter for use in the summer would require large cisterns. Because of the large amount of impervious area at airports, green roofs will likely be the most practical runoff storage method.

Green roofs, also known as ecoroofs, are a type of LID that covers a roof with vegetation (Oberndorfer et al. 2007, Dvorak and Volder 2010; see also airport applications by Velazquez 2005). There are two main types of green roofs. Extensive green roofs are shallow (< 20 cm [8 inches] of soil), with simple, low-growing plant communities that require less maintenance. Intensive green roofs have deeper soils and usually more complex plant systems; they are often referred to as rooftop gardens (Oberndorfer et al. 2007).

Controlling rooftop runoff, a component of the overall watershed area, via green roofs has a number of benefits. In addition to reducing runoff volume, the method reduces the urban heat island effect and building energy requirements (Oberndorfer et al. 2007, Dvorak and Volder 2010). Costs associated with construction of a green roof range 10–14% over conventional methods, but over the long term the annual cost can be cheaper because the vegetated environment provides a greater life cycle for the roof (40–60 years instead of the 20 years typical of a conventional roof; Carter and Keeler 2008). Essentially, the vegetation and soil provide a thermal mass that lessens wear and tear on the roof from the shrink/swell cycle (Oberndorfer et al. 2007).

A number of airports have green roofs in place. Chicago O'Hare International Airport, Chicago, Illinois, USA, for example, has found success using native grasses selected carefully to avoid wildlife attractants, and it now has >3,000 m² (32,292 feet²) of green roof on airport buildings (McAllister 2009). Native grasses were selected as ideal candidates for the control tower's vegetated roof, making it the first FAA facility of its kind in the nation. In 2010, Portland International Airport (PDX), Portland, Oregon, USA, installed a 929 m² (10,000 feet²) green roof on their new operations building (Fig. 9.3). The green roof contains 10.2-cm-deep trays with *Sedum* sp. and includes a patio area for use by employees (a component that could dissuade use by loafing birds).

We note, however, that green roofs have been proposed as potential wildlife habitat in urban areas. Brenneisen (2006) noted that the technical substrates developed for green roofs—emphasizing lightweight, consistent drainage—and efficient installation (designs compatible for use at airports) are suboptimal for biodiversity (e.g., see Brenneisen 2003). Others have noted that species richness in spiders and beetles is positively correlated with plant species richness and topographic variability in green roof designs (Oberndorfer et al. 2007). Personnel at PDX report swarms of bees when the *Sedum* sp. flowers; the bees posed no problems for operations. However, an outbreak of slugs (*Deroceras reticultatum*) on the tray-based system at PDX attracted gulls (Laridae; PDX, unpublished data); there remains the necessity to monitor performance of green roofs relative to wildlife use.

Category 4, Runoff Conveyance Practices, includes check dams, undersized culverts, and long flow paths designed to slow down and detain water for better pollutant removal, but can also create wildlife habitat, via standing water, if not properly maintained. In contrast, the long, linear nature of grassy swales might be suited for use along runways, taxiways, perimeter roads, and other paved areas (Fig. 9.4).

Category 5, Filtration Practices, includes rain gardens and vegetated swales (also described under Category 2), as well as vegetated buffers. As filtration mechanisms, however, the primary function of these methods is to remove pollutants by filtering runoff either through vegetation or by slowing flow, thereby removing suspended pollutants through settling or filtration media in the facilities (e.g., soils amended with organic or inorganic materials). Flow then enters the stormwater conveyance system rather than infiltrating to the ground, as in Category 2 approaches. When fitted with an underdrain to return flow to the conveyance system, rain gardens serve as filtration devices. Vegetated swales, also called bioswales, are vegetation-lined channels designed to remove suspended solids from stormwater. Biological uptake, biotransformation, sorption, and ion exchange are potential secondary pollutant removal processes.

Potential problems associated with filtration practices, particularly rain gardens and swales, include standing water, vegetation that attract wildlife, and weight-bearing capabilities of amended soils. Compost material is a common soil amendment because of the pollutant removal capability at relatively low cost. Where the longitudinal slope is slight, water tables are high, or continuous base flow is likely to result in saturated soil conditions, underdrains will be required to prevent standing water. Wet swales should not be used. The use of check dams across the swale to slow flows is also discouraged, as water will pool behind the dams. If flow velocities are too high through the swale, erosion can result, and the swale might need to be broken into smaller sections.

Another filtration practice, vegetated buffers (also known as vegetated filter strips), are land areas of planted vegetation and amended soils situated between the pavement surface and a surface-water collection system, basin, wetland, stream, or river. Vegetated buffers receive overland runoff from the adjacent impervious areas and rely on their flat cross slope and dense vegetation to maintain sheet flows. These buffers slow the runoff velocities, trapping sediment and other pollutants and providing some infiltration and biologic uptake.

Seattle–Tacoma International Airport (SEA), Seattle, Washington, USA, has monitored the effectiveness of vegetated buffers along their runways and taxiways and found acceptable pollutant removal within short distances (Beck and Parametrix 2006). The airport

Fig. 9.3. Green roof at Portland International Airport, Portland, Oregon, USA. Photo credit: David Felstul

Fig. 9.4. Swale and conveyance system at Denver International Airport, Denver, Colorado, USA: (A) before and (B) after improvements to the channel. Photo credit: Kendra Cross

also has added compost amendments to the soils to increase the effectiveness of pollutant removal, but the compost-amended soils attracted earthworms. If these soils are located next to paved operational areas, earthworms can invade the pavement during and after rains, providing a food source for birds (e.g., gulls). However, SEA found that using biosolids instead of compost amendments provided the high organic content for pol-

lutant removal without attracting the large numbers of worms (S. Osmek, SEA, personal communication).

Stormwater permits, such as that issued to the Port of Portland (Oregon Department of Environmental Quality 2011a), now require that LID and GI techniques be emphasized in training and in project design. In its permit fact sheet, Oregon Department of Environmental Quality (2011b) notes the critical aspect of prioritizing and incorporating LID, GI, or equivalent approaches; other program conditions such as optimizing on-site retention (i.e., infiltration, evapotranspiration, and water capture and reuse), targeting natural surface or predevelopment hydrologic functions, and minimizing hydrological and water-quality impacts of stormwater runoff from impervious surfaces.

Privately Owned Stormwater Facilities

Most public airports have large tracts of open, undeveloped land that provide added margins of safety and noise mitigation (e.g., Blackwell et al. 2009). These areas inevitably include habitats that can pose hazards to aviation, particularly if they attract wildlife to an airport's AOA or airspace. For all airports, a distance of 8 km (5 miles) between the farthest edge of the airport's AOA and the hazardous wildlife attractant is recommended if the attractant could contribute to wildlife movement into or across the approach or departure airspace (FAA 2007). However, airports and the FAA do not necessarily have control over all properties within or proximate to siting criteria. In some instances, privately owned stormwater impounds are managed for priorities that also can pose immediate hazards to aviation safety, such as general enhancement of wildlife habitat (McGuckin and Brown 1995, White and Main 2005) or use by birds for residential enjoyment, as well as biodiversity goals (Brand and Snodgrass 2009, Le Viol et al. 2009).

These contrasting priorities create a need to investigate design and management strategies that will reduce the relative attractiveness or utility of stormwater impounds to species recognized as hazardous to aviation (see Dolbeer et al. 2000, DeVault et al. 2011) while selectively targeting species (e.g., warblers, Parulidae) that pose little hazard to aviation. Specifically, impounds within or proximate to FAA siting criteria should be designed to minimize perimeter, surface area, and the ratio of emergent vegetation to open water (B. Fox, Auburn University, unpublished data). We recommend that these facilities reduce or eliminate grass areas along the pond shoreline (to reduce loafing by Canada geese) or use boulders or vegetation to break up the line of sight so as to enhance perceived predation risk (e.g., Smith et al. 1999).

Summary

Surface water composes a substantial portion (on average 6%) of U.S. airport areas within the 3-km siting criteria (B. F. Blackwell, unpublished data). Approximately 25% of bird species involved in ≥50 strikes reported to the FAA (1990–2008) have foraging and breeding ecologies closely associated with water, and over half of these species are responsible for strikes that result in aircraft damage. Research examining avian use of stormwater detention and retention ponds indicates that facility surface area, perimeter irregularity, and density of water resources within a 1-km radius are positively correlated with use by birds. Near the AOA and within or proximate to FAA siting criteria, the complete coverage of ponds physically and visually will provide the most effective means of reducing the attraction to birds. However, cover alternatives pose problems because of cost, maintenance, and water-quality issues. Both SSFW and LID/GI methods provide means of reducing peak flow, enhancing infiltration and contaminant removal, as well as reducing standing water and volume of runoff that must be contained. These methods help meet immediate safety needs for aircraft maneuvering within the AOA, while also reducing or removing water resources from wildlife use over short and long terms.

LITERATURE CITED

Airport Cooperative Research Program. 2009. Deicing planning guidelines and practices for stormwater management systems. ACRP Report 14. Transportation Research Board, Washington, D.C., USA.

Amling, W. 1980. Exclusion of gulls from reservoirs in Orange County, California. Proceedings of the Vertebrate Pest Conference 9:29–30.

Arengo, F., and G. A. Baldassarre. 2002. Patch choice and foraging behavior of nonbreeding American flamingos in Yucatan, Mexico. Condor 104:452–457.

Beck, R. W., and Parametrix. 2006. Seattle–Tacoma International Airport stormwater engineering report. Port of Seattle, Seattle, Washington, USA.

Belánger, L., and R. Couture 1988. Use of man-made ponds by dabbling duck broods. Journal of Wildlife Management 52:718–723.

Blackwell, B. F., T. L. DeVault, E. Fernandez-Juricic, and R. A. Dolbeer. 2009. Wildlife collisions with aircraft: a missing component of land-use planning for airports. Landscape and Urban Planning 93:1–9.

Blackwell, B. F., L. M. Schafer, D. A. Helon, and M. A. Linnell. 2008. Bird use of stormwater-management ponds: decreasing avian attractants on airports. Landscape and Urban Planning 86:162–170.

Blackwell, B. F., T. W. Seamans, P. M. Schmidt, T. L. DeVault, J. L. Belant, M. J. Whittingham, J. A. Martin, and E. Fernández-Juricic. 2013. A framework for managing airport grasslands and birds amidst conflicting priorities. Ibis 155:199–203.

Blokpoel, H., and G. D. Tessier. 1983. Monofilament lines exclude ring-billed gulls from traditional nesting areas. Proceedings of the Bird Control Seminar 9:15–19.

Blokpoel, H., and G. D. Tessier. 1984. Overhead wires and monofilament lines exclude ring-billed gulls from public places. Wildlife Society Bulletin 12:55–58.

Brand, A. B., and J. W. Snodgrass. 2009. Value of artificial habitats for amphibian reproduction in altered landscapes. Conservation Biology 24:295–301.

Brenneisen, S. 2003. The benefits of biodiversity from green roofs: key design consequences. Proceedings of the North American Green Roof Conference 1:323–329.

Brenneisen, S. 2006. Space for urban wildlife: designing green roofs as habitats in Switzerland. Urban Habitats 4:27–36.

Brown, M., and J. J. Dinsmore. 1986. Implications of marsh size management and isolation for marsh bird management. Journal of Wildlife Management 50:392–397.

Carter, T., and A. Keeler. 2008. Life-cycle cost-benefit analysis of extensive vegetated roof systems. Journal of Environmental Management 87:350–363.

Creighton, J. H., R. D. Sayler, J. E. Tabor, and M. J. Monda. 1997. Effects of wetland excavation on avian communities in eastern Washington. Wetlands 17:216–227.

Davies, C. M., and H. J. Bavor. 2000. The fate of stormwater-associated bacteria in constructed wetland and water pollution control pond systems. Journal of Applied Microbiology 89:349–360.

Davis, A. P. 2008. Field performance of bioretention: hydrology impacts. Journal of Hydrologic Engineering 13:90–95.

DeVault, T. L., J. L. Belant, B. F. Blackwell, and T. W. Seamans. 2011. Interspecific variation in wildlife hazards to aircraft: implications for airport wildlife management. Wildlife Society Bulletin 35:394–402.

Dietz, M. E. 2007. Low impact development practices: a review of current research and recommendations for future directions. Water, Air, and Soil Pollution 186:351–363.

Dietz, M. E., and J. C. Clausen. 2008. Stormwater runoff and export changes with development in a traditional and low impact subdivision. Journal of Environmental Management 87:560–566.

Dolbeer, R. A., P. P. Woronecki, E. C. Cleary, and E. B. Butler. 1988. Site evaluation of gull exclusion device at Fresh Kill Landfill, Staten Island, NY. Bird Damage Research Report No. 411. Denver Wildlife Research Center, Ohio Field Station, Sandusky, USA.

Dolbeer, R. A., S. E. Wright, and E. C. Cleary. 2000. Ranking the hazard level of wildlife species to aviation. Wildlife Society Bulletin 28:372–378.

Dvorak, B., and A. Volder. 2010. Green roof vegetation for North American ecoregions: a literature review. Landscape and Urban Planning 96:197–213.

FAA. Federal Aviation Administration. 2004. Title 14 U.S. Code of Federal Regulations. Part 139: certification of airports. U.S. Department of Transportation, Washington, D.C., USA.

FAA. Federal Aviation Administration. 2006. Surface drainage design. Advisory Circular 150/5320-5C. U.S. Department of Transportation, Washington, D.C., USA.

FAA. Federal Aviation Administration. 2007. Hazardous wildlife attractants on or near airports. Advisory Circular 150/5200-33B. U.S. Department of Transportation, Washington, D.C., USA.

FAA. Federal Aviation Administration. 2008. Management of airport industrial waste. Advisory Circular 150/5320-15A. U.S. Department of Transportation, Washington, D.C., USA.

FAA. Federal Aviation Administration. 2009. Airport pavement design and evaluation. Advisory Circular 15/5320-6E. U.S. Department of Transportation, Washington, D.C., USA.

FAA. Federal Aviation Administration. 2011. FAA wildlife strike database. http://wildlife-mitigation.tc.faa.gov/wildlife/default.aspx.

FAA. Federal Aviation Administration. 2012. Airport design. Advisory Circular150/5300-13A. U.S. Department of Transportation, Washington, D.C., USA.

Fairbairn, S. E., and J. J. Dinsmore. 2001. Local and landscape-level influences on wetland bird communities of the prairie pothole region of Iowa, USA. Wetlands 21:41–47.

Forsythe, D. M., and T. W. Austin. 1984. Effectiveness of an overhead wire barrier system in reducing gull use at the BFI Jedburg Sanitary Landfill, Berkeley and Dorchester counties, South Carolina. Pages 253–263 in M. J. Harrison, S. A. Gauthreaux, and L. A. Abron-Robinson, editors. Proceedings: conference and training workshop on wildlife hazards to aircraft, May 22 to May 25, 1984, Charleston, South Carolina. U.S. Department of Transportation Report DOT/FAA/AAS/84-1. Defense Technical Information Center, Washington, D.C., USA.

Gibbs, J. P., J. R. Longcore, D. G. McAuley, and J. K. Ringelman. 1991. Use of wetland habitats by selected nongame waterbirds in Maine. Publication 9. U.S. Fish and Wildlife Service, Washington, D.C., USA.

Goff, K. M., and R. W. Gentry. 2006. The influence of watershed and development characteristics on the cumulative impacts of stormwater detention ponds. Water Resources Management 20:829–860.

Higgins, J., and M. Liner. 2007. Engineering runoff solutions. Airport Business 21:22–25.

International Civil Aviation Organization. 1991. Bird control and reduction. Airport Services Manual, Document 9137-AN/898, Part 3. Montreal, Quebec, Canada.

Le Viol, I., J. Mocq, R. Julliard, and C. Kebiriou. 2009. The contribution of motorway stormwater retention ponds to the biodiversity of aquatic macroinvertebrates. Biological Conservation 42:3163–3171.

Martin, J. A., J. L. Belant, T. L. DeVault, L. W. Burger Jr., B. F. Blackwell, S. K. Riffell, and G. Wang. 2011. Wildlife risk to aviation: a multi-scale issue requires a multi-scale solution. Human–Wildlife Interactions 5:198–203.

McAllister, B. 2009. Plan green; build clean. Airport Business. January:9–11. http://www.aviationpros.com/article/10373438/plan-green-build-clean.

McAtee, W. L., and S. E. Piper. 1936. Excluding birds from reservoirs and fishponds. Leaflet No. 120. U.S. Department of Agriculture, Washington, D.C., USA.

McGuckin, C. P., and R. D. Brown. 1995. A landscape ecological model for wildlife enhancement of stormwater management practices in urban greenways. Landscape and Urban Planning 33:227–246.

McLaren, M. A., R. E. Harris, and J. W. Richardson. 1984. Effectiveness of an overhead wire barrier in deterring gulls from feeding at a sanitary landfill. Pages 241–251 in M. J. Harrison, S. A. Gauthreaux, and L. A. Abron-Robinson, editors. Proceedings: conference and training workshop on wildlife hazards to aircraft, May 22 to May 25, 1984, Charleston, South Carolina. U.S. Department of Transportation Report DOT/FAA/AAS/84-1. Defense Technical Information Center, Washington, D.C., USA.

Oberndorfer, E., J. Lundholm, B. Bass, R. R. Coffman, H. Doshi, N. Dunnett, S. Gaffin, M. Köhler, K. K. Y. Liu, and B. Rowe. 2007. Green roofs as urban ecosystems: ecological structures, functions, and services. Bioscience 57:823–833.

Oregon Department of Environmental Quality. 2011a. National Pollutant Discharge Elimination System permit: Municipal Separate Storm Sewer System (MS4) discharge permit. Issued to the City of Portland and Port of Portland. Portland, Oregon, USA. http://www.portlandoregon.gov/bes/37485.

Oregon Department of Environmental Quality. 2011b. City of Portland and Port of Portland National Pollutant Discharge Elimination System permit: Municipal Separate Storm Sewer System (MS4) permit evaluation report and fact sheet. File No. 108015. Portland, Oregon, USA. http://www.portlandoregon.gov/bes/article/322159.

Pochop, P. A., R. J. Johnson, and D. A. Aguero. 1990. The status of lines in bird damage control–review. Proceedings of the Vertebrate Pest Conference 14:317–324.

Semadeni-Davies, A. 2006. Winter performance of an urban stormwater pond in southern Sweden. Hydrological Processes 20:165–182.

Smith, A. E., S. R. Craven, and P. D. Curtis. 1999. Managing Canada geese in urban environments. Jack Berryman Institute Publication 16. Cornell University Cooperative Extension, Ithaca, New York, USA.

Suter, W. 1994. Overwintering waterfowl on Swiss lakes: how are abundance and species richness influenced by trophic status and lake morphology? Hydrobiologia 279/280:1–4.

U.S. Environmental Protection Agency. 1999. Preliminary data summary of urban stormwater best management practices. EPA-821-R-99-012. Washington, D.C., USA.

U.S. Environmental Protection Agency. 2003. When are stormwater discharges regulated as class V wells? EPA 816-F-03-001. Washington, D.C., USA.

U.S. Environmental Protection Agency. 2006. Dry detention ponds. http://cfpub.epa.gov/npdes/stormwater/menuofbmps/index.cfm?action=browse&Rbutton=detail&bmp=67.

U.S. Environmental Protection Agency. 2012. Stormwater program. http://cfpub.epa.gov/npdes/home.cfm?program_id=6.

Van de Moortel, A. M. K., D. P. L. Rousseau, F. M G. Tack, and N. De Pauw. 2009. A comparative study of surface and subsurface flow constructed wetlands for treatment of combined sewer overflows: a greenhouse experiment. Ecological Engineering 35:175–183.

Velazquez, L. S. 2005. European airport greenroofs—a potential model for North America. http://www.greenroofs.com/pdfs/exclusives-european%20_airport_greenroofs.pdf.

Washington State Department of Transportation. 2009. Aviation stormwater design manual: managing wildlife hazards near airports Seattle, Washington, USA. http://www.wsdot.wa.gov/NR/rdonlyres/587C0B2B-07B2-4D60-90DD-5E57E93F40E1/0/TableofContentsStormwater.pdf.

White, C. L., and M. B. Main. 2005. Waterbird use of created wetlands in golf-course landscapes. Wildlife Society Bulletin 33:411–421.

10

Brian E. Washburn
Thomas W. Seamans

Managing Turfgrass to Reduce Wildlife Hazards at Airports

Multiple factors—including safety regulations, economic considerations, location, and attractiveness to wildlife recognized as hazardous to aviation—influence the choice of land cover at airports. The principal land cover at airports within North America has historically been turfgrass, usually cool-season perennial grass species native to Europe. However, recent research has determined that, from a wildlife perspective, not all turfgrasses are alike. Some grasses are more palatable to herbivorous hazardous wildlife (e.g., Canada geese [*Branta canadensis*]) than others, and thus are more likely to increase the potential for wildlife–aircraft collisions when planted near critical airport operating areas. How turfgrasses are managed (e.g., by mowing or herbicide use) can also influence the degree of use by wildlife. In this chapter we (1) review the role of vegetation in the airport environment, (2) review traditional and current methods of vegetation management on airfields, (3) discuss selection criteria for plant materials in reseeding efforts, and (4) provide recommendations for future research.

Vegetation in the Airport Environment

Airports are large, complex, anthropogenically influenced environments that contain buildings and structures, impervious surfaces (e.g., pavement), and vegetated areas. Vegetation can typically be found at airports in landside areas (e.g., manicured lawns near terminal buildings and roadways), within the air op-

erations area along taxiways and runways, in larger safety areas (e.g., runway protection zones), and on airport property in outlying areas (i.e., adjacent to the airfield). Within the USA alone, airport properties include > 330,000 ha of grassland, primarily composed of areas mown at least once annually, and representing ~39–50% of airport property (DeVault et al. 2012).

Green, well-managed turfgrass represents a highly valued landscape within most societies (Ulrich 1986, Casler and Duncan 2003, Casler 2006). Turfgrass areas on the airfield and landside areas of airports add to the aesthetic—and ultimately the economic—value of the airport environment. This is particularly true when the airfield is the first part of an area seen by air travelers arriving in a new destination. An airport in a predominantly dry, desert environment might appear as an "oasis" with areas of green, growing vegetation. Managed turfgrass areas deter the amount of damage to aircraft associated with jet blast and foreign object debris, allow for the passage of aircraft straying from paved areas, and do not inhibit emergency vehicles from responding to aircraft safety incidents and accidents (Federal Aviation Administration 2011). Additionally, airfield vegetation should be relatively inflammable, tolerant to vehicle traffic and to drought, require minimal maintenance for stand persistence, prevent soil erosion, and reduce stormwater runoff. Airfield vegetation, especially near runways and taxiways, should provide limited food resources for hazardous birds (e.g., seeds, insects), provide minimal hid-

Fig. 10.1. Westover Air Reserve Base, Chicopee, Massachusetts, USA. Airfields often contain large expanses of vegetation, typically composed of turfgrasses and herbaceous vegetation. Photo credit: Brian E. Washburn

ing cover for hazardous wildlife (DeVault et al. 2011), and resist invasion by other plants that provide food and cover for wildlife (Austin-Smith and Lewis 1969, Washburn and Seamans 2004, Linnell et al. 2009; Fig. 10.1).

Numerous wildlife species (both birds and mammals) hazardous to aviation are associated with turfgrass areas at airports. In highly urbanized environments, airfields typically represent some of the largest areas of grassland habitats within those ecosystems (Kutschbach-Brohl et al. 2010, DeVault et al. 2012) and thus can be particularly attractive to hazardous birds that forage in or otherwise use open grassland habitats. Wildlife that pose a hazard to aviation use turfgrass plants and seeds directly as a food source (e.g., Canada geese, European starlings [*Sturnus vulgaris*]), or indirectly by searching for prey items such as insects and small mammals that are often found in abundance within airfield grassland habitats (e.g., raptors, coyotes [*Canis latrans*]). Other hazardous species use the open, grassland areas at airfields because the habitat conditions are similar to what these species naturally prefer (e.g., eastern meadowlarks [*Sturnella magna*], killdeer [*Charadrius vociferus*]). Wildlife use of airfield grasslands can be seasonal (e.g., during migration periods) or throughout the year, depending on the hazardous wildlife species involved, the composition of the airfield vegetation, the geographic location of the airport, and other factors.

Traditional and Current Turfgrass Management at Airfields

Traditional methods of managing wildlife habitat in grassland ecosystems, such as discing, prescribed burning, and planting food plots, benefit wildlife by providing food, cover, water, loafing areas, or other resources (Bolen and Robinson 2002). In contrast, the focus of habitat management efforts at airfields should be to develop and maintain areas that are unattractive to wildlife, particularly those species that pose a hazard to aviation (DeVault et al. 2011).

Government agencies responsible for airfield management around the world have recognized the need to manage habitat and to establish guidelines for vegetation management at airfields. Some organizations or authorities are specific in their recommendations, whereas others are vague and leave on-field management decisions to the local airport authority. The International Civil Aviation Organization (1991) recommends that grass be maintained at a height of ≥ 20 cm (8 inches). The Civil Aviation Authority (CAA) in the United Kingdom has a policy to maintain grass from 15 to 20 cm (6 to 8 inches) and describes a detailed program of mowing, thatch removal, weed control, and fertilization (CAA 2008). In the Netherlands, vegetation management involves mowing once or twice each year, followed by thatch removal within 24 hr of mowing and no fertilization of the grassland areas (Royal Netherlands Air Force 2008). This "poor grass" system reduces food for wildlife, maintains turf for erosion control, and lowers maintenance costs (Dekker and van der Zee 1996). Transport Canada does not recommend a single grass height but leaves the decision to each airport based upon the bird species that pose the greatest hazards to that airport (Transport Canada 2002). In the USA, the Federal Aviation Administration does not have a direct policy on grass height but advises airport authorities to develop vegetation management plans based on the airport's geographic location and the types of hazardous wildlife found nearby (Federal Aviation Administration 2007). The U.S. Air Force maintains a policy on grass height (U.S. Air Force Instruction 91-202, 7.11.2.3) that states, "mow airfield to maintain a uniform grass height between 7 and 14 inches [18–36 cm]." The U.S. Air Force recognizes that exceptions to this policy should occur, and a waiver procedure is available to allow for varying management practices.

The aforementioned vegetation management policies are primarily based on studies conducted in the United Kingdom in the 1970s (Brough 1971, Mead and Carter 1973, Brough and Bridgman 1980). Studies conducted at airports in the USA to determine whether tallgrass management regimens reduce bird activity have produced conflicting results (Buckley and McCarthy 1994, Seamans et al. 1999, Barras et al. 2000). These conflicting results might be due to variation in the ways studies were designed, species-specific responses of birds to vegetation height management, or variation in the density or structure of vegetation within various study locations. Seamans et al. (2007) and Washburn and Seamans (2007) found that birds exhibited species-specific responses to management practices of maintaining a set vegetation height in grasslands areas. Further, some species of birds using vegetation ≥10 cm (4 inches) tall have decreased foraging success (Baker and Brooks 1981, Whitehead et al. 1995, Atkinson et al. 2004, Devereux et al. 2004) and reduced response times to predators due to the visual obstruction of the vegetation (Bednekoff and Lima 1998, Whittingham et al. 2004, Devereux et al. 2006, Whittingham and Devereux 2008). However, flock size might be positively correlated with bird use of habitat potentially hazardous to their survival (Bednekoff and Lima 1998, Lima and Bednekoff 1999, Fernández-Juricic et al. 2004). For a given species, vegetation height and density affect escape timing and flight behavior (Devereux et al. 2008), as well as the perception of vegetation as protective cover or potential habitat for predators (Lazarus and Symonds 1992, Lima 1993). European starlings might prefer shorter vegetation but use taller vegetation when that is all that is available (Devereux et al. 2004, Seamans et al. 2007). Studies examining vegetation height seldom consider vegetation density, but when density has been presented, some bird species and numbers have decreased as vegetation density increased (Bollinger 1995, Norment et al. 1999, Scott et al. 2002, Davis 2005). Likewise, information on use of food resources (e.g. insects, seeds) in airport grasslands by birds recognized as hazardous to aviation is also limited (Bernhardt et al. 2010, Kutschbach-Brohl et al. 2010, Washburn et al. 2011). There is no published work quantifying bird response to airport grassland management that exploits visual obstruction and management of food resources to elevate perceived predation risk and reduce foraging success (Blackwell et al. 2013).

With the exception of the CAA, no organizations provide instruction in airfield vegetation height management that results in dense, uniform areas of vegetation. The CAA instructions include the use of mowing with fertilizers, which are suggested as additional limiting factors for grassland bird populations (Vickery et al. 2001). In our experience, grasslands at most North American airports present a mosaic of bare earth and plants of varying species and height, with some plants going to seed. This is the type of habitat suggested to improve conditions for survival of grassland birds in the United Kingdom (Perkins et al. 2000, Barnett et al. 2004). Also, airfields in North America as well as Europe occur within multiple ecosystems, resulting in a wide range of vegetative conditions (i.e., plant communities) and wildlife issues unique to each area. More research is needed to fully understand how vegetation height, combined with other plant community characteristics (e.g., plant species composition, vegetation density), influences the use of airfield habitats by hazardous birds (Washburn and Seamans 2007; Blackwell et al. 2009, 2013). When developing a vegetation management plan for a specific airport, airfield managers and wildlife biologists should consider the species of hazardous wildlife that typically use the airfield and manage vegetation height to minimize use by those particular species (DeVault et al. 2011).

Other tools, such as herbicides and plant growth regulators, might also be applied to control or manage airfield vegetation (Blackwell et al. 1999, Washburn and Seamans 2007, Washburn et al. 2011). Depending on the vegetation management objectives desired, herbicides allow managers to alter vegetative composition of the airfield by removing (e.g., killing) or favoring (e.g., by removing competition) certain types of plants. Broadleaf-selective herbicides, such as 2,4-D or dicamba, could be used to remove broad-leaved forbs and legumes (e.g., clovers [*Trifolium* spp.]) that are an attractive food resource for certain wildlife species, including herbivorous birds (e.g., Canada geese) and cervids (e.g., white-tailed deer [*Odocoileus virginianus*]).

In addition, grassland plant communities with lower structural and botanical diversity support insect populations with less abundance and diversity (Tscharntke and Greiler 1995, Morris 2000), reducing potential food sources for insectivorous birds. Selective herbicide applications on airfields represent a management option for creating a near monoculture of a desired plant (e.g., seeded turfgrasses) and reducing foraging opportunities for wildlife hazardous to aviation (Washburn and Seamans 2007, Washburn et al. 2011).

Plant growth regulators are commonly used in traditional turfgrass management to reduce growth of vegetation and maintain vegetation at a desired height (Christians 2011). In situations where plant growth regulators might control the growth of vegetation (e.g., turfgrass) as an alternative to mowing, the use of such chemicals may not be cost-effective when compared to mowing only (Washburn and Seamans 2007). Some airports are limited to the number and type of pesticide (chemicals) that can be applied to the airfield; this is especially true of U.S. Department of Defense airfields and installations. Specialized equipment and application licenses also may be required before chemicals can be stored on site or applied to airfield vegetation.

Airfield grassland habitats often contain diverse insect communities that represent a potential attractant to hazardous birds (Kutschbach-Brohl et al. 2010). Laughing gulls (*Larus atricilla*) use airfield habitats for foraging and prey upon Japanese beetles (*Popillia japonica*) and other insects as a seasonal food source (Buckley and McCarthy 1994, Caccamise et al. 1994, Bernhardt et al. 2010). Garland et al. (2009) and Washburn et al. (2011) documented that American kestrels (*Falco sparverious*) struck by aircraft or collected during wildlife control operations at North American airports had eaten grasshoppers. Management of insect pest populations at airports (i.e., turf-damaging true bugs [Hemiptera], grasshoppers [Orthoptera], and beetles [Coleoptera]) identified as a food source (and therefore an attractant to hazardous wildlife) provides an opportunity to reduce the risk of wildlife–aircraft collisions (Washburn et al. 2011). Dietary information from hazardous species using airports (Chapter 8) could be particularly useful in developing effective management options. A variety of insecticides, commonly used to control turf pests, are commercially available for use on airfield grasslands and turfgrass areas (Christians 2011).

Populations of small mammals inhabiting airfield grassland habitats can represent an important attractant (as prey items) for hazardous birds (e.g., hawks and owls) and mammals (e.g., coyotes). Large or dense populations of small mammals at airports can result in increased risk of wildlife–aircraft collisions by bird and mammal predators (see Chapter 8 for a discussion of the confounding issue of abundance versus availability of small mammals to predators). Small mammals can be managed most effectively through an integrated pest management approach, which might involve vegetation management (e.g., mowing), applying toxic rodenticide baits (e.g., zinc phosphide), altering plant communities, or combining various methods (Witmer and Fantinato 2003, Witmer 2011). Numerous studies demonstrate that mowing vegetation reduces small-mammal use of grassland habitats (Grimm and Yahner 1988, Edge et al. 1995, Seamans et al. 2007, Washburn and Seamans 2007).

Prescribed fire (i.e., burning) is one of the most widely used and effective tools for managing grassland habitats (Packard and Mutel 1997, Washburn et al. 2000). Benefits from prescribed fire used within grasslands include removal of encroaching woody vegetation, release of nutrients, and enhancement of native plant populations. Given safety issues related to reduced visibility from smoke and other hazards associated with prescribed burning, this technique has not been frequently used within airport environments. However, burning has potential for assisting with airfield vegetation management goals in situations where the related safety issues could be managed effectively.

Airfield grasslands represent a potential location for biosolids application and a possible source of revenue for airports. Application of municipal biosolids (i.e., treated and stabilized sewage sludge) provides plant nutrition and organic matter inputs to receiving grassland systems. At one military airfield in North Carolina, Washburn and Begier (2011) found that although long-term biosolids applications altered plant communities, the relative hazards posed by wildlife using treated and untreated areas were similar.

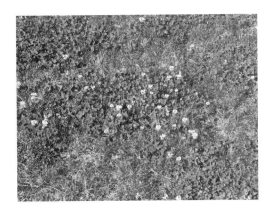

Fig. 10.2. Grass and clover (*Trifolium* spp.). Plant communities on airfields are often diverse and contain plants, such as these, which are selected by hazardous wildlife species. Photo credit: Brian E. Washburn

Selecting Plant Material for Renovation and Initial Seeding Efforts

Species composition of plant communities (the types of plants) in airfield areas also has the potential to impact the degree of attractiveness of airfields to hazardous wildlife and other bird attractants (e.g., insects, small mammals; Austin-Smith and Lewis 1969, Dekker and van der Zee 1996, Washburn and Seamans 2007). Selecting plant materials (e.g., turfgrasses) that are minimally attractive to hazardous wildlife and avoiding those that are commonly utilized by hazardous birds and mammals are important decisions, and present an important management opportunity for revegetation projects at airports (Fig. 10.2).

Abiotic factors (e.g., climatic conditions, soil nutrient levels) and biotic factors (e.g., weed competition) have strong influence on the rate of establishment of turfgrasses and other plants seeded as part of an airfield renovation or revegetation project. These abiotic and biotic factors can vary greatly among airports, depending on the airport's geographic location, local geology and soils, and regional climatic conditions. Some factors, such as weather, cannot be controlled or predicted and are not under control of airfield managers. In contrast, other factors can be monitored and amended when establishing turfgrasses on airfields by using methods such as soil testing and fertilization, planting good-quality turfgrass seed, and applying ap-

propriate chemical control (e.g., herbicides) to reduce weed competition (Willis et al. 2006). Washburn et al. (2007b) found that tall fescue (*Schedonorus phoenix*) establishment on airfields was enhanced by the application of mulch during seeding efforts. Specific guidelines for seeding and sodding turfgrasses at airports are available (see Federal Aviation Administration 2009). In addition, many airports and state aviation departments have developed their own airport- and region-specific specifications for seeding and sodding areas within airport environments.

Cover crops are commonly used in seeding efforts following construction projects because they quickly establish vegetative cover to prevent erosion. Use of cover crops, which typically consist of cereal grains (e.g., browntop millet [*Panicum ramosum*], wheat [*Triticum* spp.], and cereal rye [*Secale cereale*]), should not be used for reseeding efforts in airport environments, as they provide food for granivorous birds (e.g., mourning doves [*Zenaida macroura*] and rock pigeons [*Columba livia*]) and other birds recognized as hazardous to aviation (Castellano 1998).

Tall fescue is a cool-season, perennial sod-forming grass that grows well in temperate regions. In recent years, this traditionally agronomic forage grass has been developed as a turfgrass and has become very popular for use in parks, lawns, golf courses, sports fields, and other areas (Casler 2006). Tall fescue is frequently infested with a fungal endophyte (*Neotyphodium coenophialum*) that forms a mutualistic, symbiotic relationship with the grass. Grasses containing endophytic fungi derive several benefits, such as resistance to both grazing and insect herbivory, increased heat and drought stress tolerance, and increased vigor (Clay et al. 1985, Vicari and Bazely 1993, Malinowski and Belesky 2000). Tall fescue is also extremely competitive and develops into solid stands, crowding out other grasses, legumes, and annual weeds (Barnes et al. 1995, Richmond et al. 2006, Washburn et al. 2007a). Tall fescue grasslands are generally unattractive as a foraging resource to grassland wildlife (Barnes et al. 1995, Washburn et al. 2000), making it a potentially useful ground cover for airfields in some regions of the USA.

The turfgrass industry has recently developed a large number of "turf-type" cultivars of the most common varieties (e.g., tall fescue, perennial ryegrass [*Lo-*

lium perenne], bermudagrass [*Cynodon dactylon*]) for lawns, golf courses, parks, and other traditional uses. Turf-type cultivars are bred for horticultural characteristics important to the turfgrass industry, namely deep green color, drought and disease resistance, and shorter growth habits than traditional agronomic tall fescue cultivars. In addition, many of the new tall fescue cultivars have high levels of *Neotyphonium* endophyte infection (Mohr et al. 2002). Consequently, a diverse variety of plant materials are commercially available for use in airfield and landside seeding and revegetation projects within airport environments. Washburn et al. (2007b) found that high-endophyte turf-type tall fescue cultivars did establish on areas within airports.

The various species of commercially available turfgrass commonly used in North America differ in their attractiveness and utility as food for wildlife hazardous to aviation. Washburn and Seamans (2012) found that Canada geese exhibited clear preferences among commercially available turfgrasses when foraging. Their findings suggest that selected commercial turfgrasses (e.g., zoysiagrass [*Zoysia japonica*], centipedegrass [*Eremochloa ophiuroides*], and St. Augustinegrass [*Stenotaphrum secundatum*]) might be preferable in reseeding and vegetation renovation projects within areas where Canada geese are unwanted (e.g., airfields, parks, athletic fields, and golf courses). In contrast, creeping bentgrass (*Agrostis stolonifera*), Kentucky bluegrass (*Poa pratensis*), and fine fescues (*Festuca* spp.) were favored by foraging geese and thus should be avoided when formulating seed mixtures for reseeding areas at airfields and other places where human–goose conflicts might occur (Fig. 10.3).

Grazing Anatidae (including various species of geese) apparently make foraging choices based on the nutritional content and chemical composition of plants (Gauthier and Bedard 1991, McKay et al. 2001, Durant et al. 2004). The concentration of protein within forage plants is an important component in the foraging preferences for a variety of goose species, including barnacle geese (*B. leucopsis*; Prins and Ydenberg 1985), dark-bellied brent geese (*B. bernicla bernicla*; McKay et al. 2001), white-fronted geese (*Anser albifrons albifrons*; Owen 1976), graylag geese (*A. anser*; Van Liere et al. 2009), and Canada geese (Sedinger and Raveling 1984, Washburn and Seamans 2012). In addition, sec-

Fig. 10.3. Canada geese in study plot. Research examining the foraging preferences of Canada geese provides information about the selection of plant materials for reseeding projects at airports. Photo credit: Brian E. Washburn

ondary plant defense compounds, such as alkaloids and tannins, can cause geese to avoid certain plants. Physical characteristics of plants (e.g., leaf tensile strength, hairy leaves) also might influence whether geese (or other wildlife) forage on specific plants (Lieff et al. 1970, Williams and Forbes 1980, Buchsbaum et al. 1984, Conover 1991, Gauthier and Hughes 1995).

In addition to the commercially available turfgrasses, other plants could be appropriate for use on airfields. Linnell et al. (2009) found that wedelia (*Wedelia trilobata*), a mat-forming composite plant, showed promise for use as a vegetative cover on tropical airfields. Pochop et al. (1999) found several grasses that were locally adapted to Alaska and not favored by Canada geese in captive situations. These studies are limited to specific ecotypes and locations but demonstrate the availability of regionally specific plants that are not desired by wildlife hazardous to aviation. Some caution is warranted when establishing nonnative plant species: doing so can result in the escape of exotic species from the airfield environment (Austin-Smith and Lewis 1969). We recommend consulting state departments of agriculture or natural resources Web sites for lists of invasive plants that are illegal to plant because of their noxious or nonnative status. Native plants that are aesthetically pleasing, legal to plant, and unattractive to hazardous wildlife are the best choices for airfield vegetation, assuming seed is available in sufficient quantities and at reasonable cost.

Summary

Although some research has been conducted on the use of airfield grasslands by wildlife hazardous to aviation, much is still unknown. Field research within active airport environments to determine the most effective tools and techniques to reduce insect resources and small mammal populations, and consequently hazardous wildlife use of airfields, is needed. Research examining the use of prescribed fire, herbicides, growth regulators, and other airfield vegetation management methods should also be a priority. As with other aspects of managing human–wildlife conflicts in the airport environment, we suggest that future studies focus on those species that are most hazardous to aircraft (Dolbeer et al. 2000, 2010; DeVault et al. 2011). With regard to airfield grasslands, species that use these habitats for foraging and to meet other life history needs should be of a high priority. Information about the influence of unique land uses (such as biosolids disposal and biofuel production; Chapter 11) on the attractiveness of airfields to hazardous wildlife represents an important area for future research, making it beneficial in the formulation and evaluation of effective management practices for airport environments.

LITERATURE CITED

Atkinson, P. W., D. Buckingham, and A. J. Morris. 2004. What factors determine where invertebrate-feeding birds forage in dry agricultural grasslands? Ibis 146:99–107.

Austin-Smith, P. J., and H. F. Lewis. 1969. Alternative vegetative ground cover. Pages 153–160 *in* Proceedings of the world conference on bird hazards to aircraft, 2–5 September 1969. National Research Council of Canada, Kingston, Ontario, Canada.

Baker, J. A., and R. J. Brooks. 1981. Raptor and vole populations at an airport. Journal of Wildlife Management 45:390–396.

Barnes, T. G., L. A. Madison, J. D. Sole, and M. J. Lacki. 1995. An assessment of habitat quality for northern bobwhite in tall fescue–dominated fields. Wildlife Society Bulletin 23:231–237.

Barnett, P. R., M. J. Whittingham, R. B. Bradbury, and J. D. Wilson. 2004. Use of unimproved and improved lowland grassland by wintering birds in the UK. Agriculture, Ecosystems, and Environment 102:49–60.

Barras, S. C., R. A. Dolbeer, R. B. Chipman, G. E. Bernhardt, and M. S. Carrara. 2000. Bird and small mammal use of mowed and unmowed vegetation at John F. Kennedy International Airport, 1998–1999. Proceedings of the Vertebrate Pest Conference 19:31–36.

Bednekoff, P. A., and S. L. Lima. 1998. Re-examining safety in numbers: interactions between risk dilution and collective detection depend upon predator targeting behaviour. Proceedings of the Royal Society B 265:2021–2026.

Bernhardt, G. E., L. Kutschbach-Brohl, B. E. Washburn, R. B. Chipman, and L. C. Francoeur. 2010. Temporal variation in terrestrial invertebrate consumption by laughing gulls in New York. American Midland Naturalist 163:442–454.

Blackwell, B. F., T. L. DeVault, E. Fernández-Juricic, and R. A. Dolbeer. 2009. Wildlife collisions with aircraft: a missing component of land-use planning on and near airports? Landscape and Urban Planning 93:1–9.

Blackwell, B. F., T. W. Seamans, and R. A. Dolbeer. 1999. Plant growth regulator (Stronghold™) enhances repellency of anthraquinone formulation (Flight Control™) to Canada geese. Journal of Wildlife Management 63:1336–1343.

Blackwell, B. F., T. W. Seamans, P. M. Schmidt, T. L. DeVault, J. L. Belant, M. J. Whittingham, J. A. Martin, and E. Fernández-Juricic. 2013. A framework for managing airport grasslands and birds amidst conflicting priorities. Ibis 155:199–203.

Bolen, E. G., and W. L. Robinson. 2002. Wildlife ecology and management. Fifth edition. Prentice Hall, New York, New York, USA.

Bollinger, E. K. 1995. Successional changes and habitat selection in hayfield bird communities. Auk 112:720–730.

Brough, T. 1971. Experimental use of long-grass in the U.K. Proceedings of the Bird Strike Committee Europe 6.

Brough, T., and C. J. Bridgman. 1980. An evaluation of long grass as a bird deterrent on British airfields. Journal of Applied Ecology 17:243–253.

Buchsbaum, R., I. Valiela, and T. Swain. 1984. The role of phenolic compounds and other plant constituents in feeding by Canada geese in a coastal marsh. Oecologia 63:343–349.

Buckley, P. A., and M. G. McCarthy. 1994. Insects, vegetation, and the control of laughing gulls (*Larus atricilla*) at Kennedy International Airport, New York City. Journal of Applied Ecology 31:291–302.

CAA. Civil Aviation Authority. 2008. CAP 772: birdstrike risk management for aerodromes. Safety Regulation Group, Gatwick Airport South, West Sussex, United Kingdom.

Caccamise, D. F., J. J. Dosch, K. Bennett, L. M. Reed, and L. DeLay. 1994. Management of bird strike hazards at airports: a habitat approach. Proceedings of the Bird Strike Committee Europe 22:285–306.

Casler, M. D. 2006. Perennial grasses for turf, sport, and amenity uses: evolution of form, function, and fitness for human benefit. Journal of Agricultural Science 144:189–203.

Casler, M. D., and R. R. Duncan. 2003. Turfgrass biology, genetics, and breeding. John Wiley and Sons, New York, New York, USA.

Castellano, B. 1998. Grasses attractive to hazardous wildlife. CertAlert 98-05. Federal Aviation Administration, Airports Safety and Operations Division, Washington, D.C., USA.

Christians, N. 2011. Fundamentals of turfgrass management. Fourth edition. John Wiley and Sons, New York, New York, USA.

Clay, K., T. N. Hardy, and A. M. Hammond Jr. 1985. Fungal endophytes of grasses and their effects on an insect herbivore. Oecologia 66:1–5.

Conover, M. R. 1991. Herbivory by Canada geese: diet selection and effect on lawns. Ecological Applications 1:231–236.

Davis, S. K. 2005. Nest-site selection patterns and the influence of vegetation on nest survival of mixed-grass prairie passerines. Condor 107:605–616.

Dekker, A., and F. F. van der Zee. 1996. Birds and grasslands on airports. Proceedings of the International Bird Strike Committee 23:291–305.

DeVault, T. L., J. L. Belant, B. F. Blackwell, J. A. Martin, J. A. Schmidt, and L. W. Burger Jr. 2012. Airports offer unrealized potential for alternative energy production. Environmental Management 49:517–522.

DeVault, T. L., J. L. Belant, B. F. Blackwell, and T. W. Seamans. 2011. Interspecific variation in wildlife hazards to aircraft: implications for airport wildlife management. Wildlife Society Bulletin 35:394–402.

Devereux, C. L., E. Fernández-Juricic, J. R. Krebs, and M. J. Whittingham. 2008. Habitat affects escape behaviour and alarm calling in Common Starlings Sturnus vulgaris. Ibis 150:191–198.

Devereux, C. L., C. U. McKeever, T. G. Benton, and M. J. Whittingham. 2004. The effect of sward height and drainage on Common Starlings Sturnus vulgaris and Northern Lapwings Vanellus vanellus foraging in grassland habitats. Ibis 146:115–122.

Devereux, C. L., M. J. Whittingham, E. Fernández-Juricic, J. A. Vickery, and J. R. Krebs. 2006. Predator detection and avoidance by starlings under differing scenarios of predation risk. Behavioural Ecology 17:303–309.

Dolbeer, R. A., S. E. Wright, and E. C. Cleary. 2000. Ranking the hazard level of wildlife species to aviation. Wildlife Society Bulletin 28:372–378.

Dolbeer, R. A., S. E. Wright, J. Weller, and M. J. Begier. 2010. Wildlife strikes to civil aircraft in the United States 1990–2009. Serial Report 16. U.S. Department of Transportation, Federal Aviation Administration, Office of Airport Safety and Standards, Washington, D.C., USA.

Durant, D., H. Fritz, and P. Duncan. 2004. Feeding patch selection by herbivorous Anatidae: the influence of body size, and of plant quantity and quality. Journal of Avian Biology 25:144–152.

Edge, W. D., J. O. Wolff, and R. L. Carey. 1995. Density-dependent responses of gray-tailed voles to mowing. Journal of Wildlife Management 59:245–251.

Federal Aviation Administration. 2007. Hazardous wildlife attractants on or near airports. Advisory Circular 150/5200-33B. U.S. Department of Transportation, Washington, D.C., USA.

Federal Aviation Administration. 2009. Standards for specifying construction of airports. Advisory Circular 150/5370-10E. U.S. Department of Transportation, Washington, D.C., USA.

Federal Aviation Administration. 2011. Airport planning. Advisory Circular 150/5300-13. U.S. Department of Transportation, Washington, D.C., USA.

Fernández-Juricic, E., S. Siller, and A. Kacelnik. 2004. Flock density, social foraging, and scanning: an experiment with starlings. Behavioural Ecology 15:371–379.

Garland, H., A. D. Peral, D. M. Bird, and M. A. Fortin. 2009. High frequency of American kestrel strikes at the Montreal–Trudeau International Airport: a case study. Journal of Raptor Research 43:382–383.

Gauthier, G., and J. Bedard. 1991. Experimental tests of the palatability of forage plants in greater snow geese. Journal of Applied Ecology 28:491–500.

Gauthier, G., and R. J. Hughes. 1995. The palatability of Arctic willow for greater snow geese: the role of nutrients and deterring factors. Oecologia 103:390–392.

Grimm, J. W., and R. H. Yahner. 1988. Small mammal response to roadside habitat management in central Minnesota. Journal of the Minnesota Academy of Science 53:16–21.

International Civil Aviation Organization. 1991. Bird control and reduction. Airport Services Manual, Document 9137 -AN/898, Part 3. Montreal, Quebec, Canada.

Kutschbach-Brohl, L., B. E. Washburn, G. E. Bernhardt, R. B. Chipman, and L. C. Francoeur. 2010. Arthropods of a semi-natural grassland in an urban environment: the John F. Kennedy International Airport, New York. Journal of Insect Conservation 14:347–358.

Lazarus, J., and M. Symonds. 1992. Contrasting effects of protective and obstructive cover on avian vigilance. Animal Behaviour 43:519–521.

Lieff, B. C., C. D. MacInnes, and R. K. Kistra. 1970. Food selection experiments with young geese. Journal of Wildlife Management 34:321–327.

Lima, S. L. 1993. Ecological and evolutionary perspectives on escape from predatory attack: a survey of North American birds. Wilson Bulletin 105:1–47.

Lima, S. L., and P. A. Bednekoff. 1999. Back to the basics of antipredatory vigilance: can nonvigilant animals detect attack? Animal Behaviour 58:537–543.

Linnell, M. A., M. R. Conover, and T. J. Ohashi. 2009. Using wedelia as ground cover on tropical airports to reduce bird activity. Human–Wildlife Conflicts 3:226–236.

Malinowski, D. P., and D. P. Belesky. 2000. Adaptations of endophyte-infected cool-season grasses to environmental stresses: mechanisms of drought and mineral stress tolerance. Crop Science 40:923–940.

McKay, H. V., T. P. Milsom, C. J. Feare, D. C. Ennis, D. P. O'Connell, and D. J. Haskell. 2001. Selection of forage species and the creation of alternative feeding areas for dark-bellied brent geese (*Branta bernicla bernicla*) in southern UK coastal areas. Agriculture, Ecosystems, and Environment 84:99–113.

Mead, H., and A. W. Carter. 1973. The management of long grass as a bird repellent on airfields. Journal of the British Grassland Society 28:219–221.

Mohr, M. M., W. A. Meyer, and C. Mansue. 2002. Incidence of *Neotyphodium* endophyte in seed lots of cultivars and selections of the 2001 national tall fescue test. Rutgers Turfgrass Proceedings 34:196–201.

Morris, M. G. 2000. The effects of structure and its dynamics on the ecology and conservation of arthropods in British grasslands. Biological Conservation 95:129–142.

Norment, C. J., C. D. Ardizzone, and K. Hartman. 1999. Habitat relations and breeding biology of grassland birds in New York. Studies in Avian Biology 19:112–121.

Owen, M. 1976. The selection of winter food by white-fronted geese. Journal of Applied Ecology 13:715–729.

Packard, S., and C. F. Mutel. 1997. The tallgrass restoration handbook for prairies, savannas, and woodlands. Island Press, Washington, D.C., USA.

Perkins, A. J., M. J. Whittingham, R. B. Bradbury, J. D. Wilson, A. J. Morris, and P. R. Barnett. 2000. Habitat characteristics affecting use of lowland agricultural grassland by birds in winter. Biological Conservation 95:279–294.

Pochop, P. A., J. L. Cummings, K. L. Wedemeyer, R. M. Engeman, and J. E. Davis Jr. 1999. Vegetation preferences of captive Canada geese at Elmendorf Air Force Base, Alaska. Wildlife Society Bulletin 27:734–740.

Prins, H. H. T., and R. C. Ydenberg. 1985. Vegetation growth and a seasonal habitat shift of the barnacle goose (*Branta leucopsis*). Oecologia 66:122–125.

Richmond, D. S., J. Cardina, and P. S. Grewal. 2006. Influence of grass species and endophyte infection on weed populations during establishment of low maintenance lawns. Agriculture, Ecosystems, and Environment 115:27–33.

Royal Netherlands Air Force. 2008. Voorschrift Vogelaanvaringspreventie (1ᴱ Uitgave). Behandelende Instantie: CLSK/DO/HAMO. Publication No. 063178. Breda, Netherlands.

Scott, P. E., T. L. DeVault, R. A. Bajema, and S. L. Lima. 2002. Grassland vegetation and bird abundances on reclaimed Midwestern coal mines. Wildlife Society Bulletin 30:1006–1014.

Seamans, T. W., S. C. Barras, G. E. Bernhardt, B. F. Blackwell, and J. D. Cepek. 2007. Comparison of 2 vegetation-height management practices for wildlife control at airports. Human–Wildlife Conflicts 1:97–105.

Seamans, T. W., R. A. Dolbeer, M. S. Carrara, and R. B. Chipman. 1999. Does tall grass reduce bird numbers on airports? Results of pen test with Canada geese and field trials at two airports, 1998. Pages 160–170 in Proceedings of the first joint annual meeting. Bird Strike Committee–USA/Canada. Vancouver, British Columbia, Canada.

Sedinger, J. S., and D. G. Raveling. 1984. Dietary selectivity in relation to availability and quality of food for goslings of cackling geese. Auk 101:295–306.

Transport Canada. 2002. Wildlife control procedures manual. TP 11500 E. Third edition. Transport Canada, Safety and Security, Aerodrome Safety Branch, Ottawa, Ontario, Canada.

Tscharntke, T., and H. J. Greiler. 1995. Insect communities, grasses, and grasslands. Annual Review of Entomology 40:535–558.

Ulrich, R. S. 1986. Human response to vegetation and landscapes. Landscape and Urban Planning 13:29–44.

Van Liere, D. W., N. J. M. Van Eekeren, and M. J. J. E. Loonen. 2009. Feeding preferences of greylag geese and the effect of activated charcoal. Journal of Wildlife Management 73:924–931.

Vicari, M., and D. R. Bazely. 1993. Do grasses fight back? The case for antiherbivore defenses. Trends in Ecology and Evolution 8:137–141.

Vickery, J. A., J. R. Tallowin, R. E. Feber, E. J. Asteraki, P. W. Atkinson, R. J. Fuller, and V. K. Brown. 2001. The management of lowland neutral grasslands in Britain: effects of agricultural practices on birds and their food resources. Journal of Applied Ecology 38:647–664.

Washburn, B. E., T. G. Barnes, and J. D. Sole. 2000. Improving northern bobwhite habitat by converting tall fescue fields to native warm-season grasses. Wildlife Society Bulletin 28:97–104.

Washburn, B. E., S. C. Barras, and T. W. Seamans. 2007a. Foraging preferences of captive Canada geese related to turfgrass mixtures. Human–Wildlife Conflicts 1:214–223.

Washburn, B. E., and M. J. Begier. 2011. Wildlife responses to long-term application of biosolids to grasslands in North Carolina. Rangeland Ecology and Management 64:131–138.

Washburn, B. E., G. E. Bernhardt, and L. A. Kutschbach-Brohl. 2011. Using dietary analyses to reduce the risk of wildlife–aircraft collisions. Human–Wildlife Interactions 5:204–209.

Washburn, B. E., J. S. Loven, M. J. Begier, D. P. Sullivan, and H. A. Woods. 2007b. Evaluating commercially available tall fescue varieties for airfields. Pages 1–12 in Proceedings of the FAA worldwide airport technology transfer conference. Federal Aviation Administration, Washington, D.C., USA.

Washburn, B. E., and T. W. Seamans. 2004. Management of vegetation to reduce wildlife hazards at airports. Pages 1–7 in Proceedings of the FAA worldwide airport technology transfer conference. Federal Aviation Administration, Washington, D.C., USA.

Washburn, B. E., and T. W. Seamans. 2007. Wildlife responses to vegetation height management in cool-season grasslands. Rangeland Ecology and Management 60:319–323.

Washburn, B. E., and T. W. Seamans. 2012. Foraging preferences of Canada geese: implications for reducing human–goose conflicts. Journal of Wildlife Management 76:600–607.

Whitehead, S. C., J. Wright, and P. A. Cotton. 1995. Winter field use by the European starling *Sturnus vulgaris*: habitat preferences and the availability of prey. Journal of Avian Biology 26:193–202.

Whittingham, M. J., S. J. Butler, J. L. Quinn, and W. Cress-well. 2004. The effect of limited visibility on vigilance behaviour and speed of predator detection: implications for the conservation of granivorous passerines. Oikos 106:377–385.

Whittingham, M. J., and C. L. Devereux. 2008. Changing grass height alters foraging site selection by wintering farmland birds. Basic and Applied Ecology 9:779–788.

Williams, G., and J. E. Forbes. 1980. The habitat and dietary preferences of dark-bellied brent geese and wigeon in relation to agricultural management. Wildfowl 31:151–157.

Willis, J. B., J. B. Beam, W. L. Barker, and S. D. Askew. 2006. Weed control options in spring-seeded tall fescue (*Festuca arundinacea*). Weed Technology 20:1040–1046.

Witmer, G. W. 2011. Rodent population management at Kansas City International Airport. Human–Wildlife Interactions 5:269–275.

Witmer, G. W., and J. W. Fantinato. 2003. Management of rodent populations at airports. Proceedings of the Wildlife Damage Management Conference 10:350–357.

11

JAMES A. MARTIN
TARA J. CONKLING
JERROLD L. BELANT
KRISTIN M. BIONDI
BRADLEY F. BLACKWELL
TRAVIS L. DEVAULT
ESTEBAN
FERNÁNDEZ-JURICIC
PAIGE M. SCHMIDT
THOMAS W. SEAMANS

Wildlife Conservation and Alternative Land Uses at Airports

Given all the attention paid throughout this book to minimizing the risk of wildlife–aircraft strikes, the title of this chapter may seem like an oxymoron. This book has emphasized management as related to the hazardous (to aircraft) sector of biodiversity. In this chapter we focus on the issue of protection and management of less hazardous taxa, and how altering land use at airports might, in limited circumstances, contribute to this objective.

The term "conservation" often leads to confusion and perceived conflicting goals of management. In fact, many of the direct management techniques used at airports (e.g., deterrents, translocation, etc.) could be considered conservation measures, because they remove birds from harm's way. None of these techniques are designed to extirpate a species from the environment; they are employed to reduce or remove risk to aviation, as well as the birds themselves (Blokpoel 1976, Conover 2002). Even in cases where lethal population control is used, the species involved are typically common and not threatened with extinction. In the context of this chapter we define conservation as the "protection and management of biodiversity" (Groom et al. 2006).

Conservation biologists and other scientists have debated whether wildlife conservation, such as promoting grassland birds, is an appropriate objective for airports (Kelly and Allan 2006, Blackwell et al. 2013). However, there is a lack of scientific literature on this topic to provide the necessary guidance. The ambiguity of promoting conservation at airports exists because of numerous factors, including imperfect information about wildlife response to habitat management or altering land use, variation in human values for certain wildlife taxa, and spatial variations in wildlife resource needs. Research based on ecological and animal behavior principles is necessary to achieve a safe airport environment while having any hope for wildlife conservation (Blackwell et al. 2013). Nevertheless, wildlife management at airports must continue in the face of uncertainty. Our goal is to provide background information necessary to reduce ambiguity on this issue as well as a roadmap for consideration of future conservation and applied research efforts.

Current Land Use and Implications for Wildlife

The connections between land use, land cover, and wildlife habitat are at the forefront of conserving wildlife at airports (Blackwell et al. 2009). Land use can be defined as how and why humans employ the land and its resources (Meyer 1995, Turner et al. 2001). Land cover refers to the "vegetation type present such as forest, agriculture, and grassland" (Turner et al. 2001). We use Hall et al.'s (1997) definition of habitat as "the resources and conditions present in an area that produce occupancy—including survival and reproduction—by a given organism." In the context of the airport environment, most species' habitat requirements will not be met solely on airport property, requiring movements to and from the airport (which, incidentally, could increase strike risk; Chapter 12). The airport proper may be used for specific resource needs, such as food (Chapter 8). For some grassland species, however, seasonal habitat may exist only on airport property (Kershner and Bollinger 1996). Eastern meadowlarks (*Sturnella magna*) are grassland-obligate birds that forage and nest in grass-dominated areas (e.g., hayfields or mowed airport fields; Roseberry and Klimstra 1970), whereas European starlings (*Sturnus vulgaris*) are a facultative-grassland species that forage in grasslands but nest in cavities (Kessel 1957). Meadowlarks require only a single land use or cover type; starlings minimally require two land-use/cover types to fulfill their life history requirements. Not only does this simple example demonstrate the importance of terminology usage, but it has important implications for management. Control or conservation of meadowlarks could conceivably be achieved in a single grassland patch within the airport boundary. However, management of starlings to reduce use at the airport may require alterations of two land-use types—mowed fields and structures offering cavities—making the task more difficult.

Wildlife occupancy of various land-use/cover types can markedly influence the risk of wildlife collisions with aircraft. The International Civil Aviation Organization (2002) provides this summary of the effects of certain land uses on wildlife hazards:

> Land uses considered as contributing to wildlife hazards on or near [i.e., within 13 km] airports are fish-processing operations; agriculture; livestock feed lots; refuse dumps and landfills; factory roofs; parking lots; theaters and food outlets; wildlife refuges; artificial and natural lakes; golf and polo courses, etc.; animal farms; and slaughter houses.

In addition, the International Civil Aviation Organization grades land uses as to whether they are acceptable within radii from the airport center of 3 and 8 km (1.9 and 5 miles). The Federal Aviation Administration (2007) also provides guidance for hazardous attractants at or near airports. Other chapters in this book discuss land-use/cover types, including water resources (Chapter 9), turfgrass (a form of grassland; Chapter 10), and trash facilities (included in Chapter 8). These land-use/cover types can represent a substantial portion of the area surrounding airports; other land uses may include agriculture as well as alternative

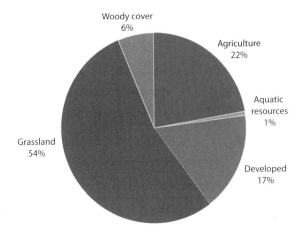

Fig. 11.1. Percentage of land cover or habitat type for 10 small airports in Indiana, USA. Adapted from DeVault et al. (2009)

energy crops and sources (DeVault et al. 2009, 2012). In this chapter we briefly discuss agriculture, including alternative energy crops, and its value for avian conservation and hazardous species reduction, as well as habitat needs of grassland birds.

Agriculture as a Land Use, Cover Type, and Habitat Component

As noted above, airports consist of a wide range of land cover and potential habitat types (Fahrig 2003, DeVault et al. 2009; Fig. 11.1). The degree to which habitat contributes to wildlife–aircraft strike risk at airports should not be based on the overall number of wildlife species that use the cover, however, but on the relative hazards those species pose to aircraft (DeVault et al. 2011). A land cover with greater wildlife abundance and diversity may actually represent a lower hazard to aircraft and might be more suitable for use at airports. Robertson et al. (2011) compared bird communities in three different land covers, including corn (*Zea mays*), switchgrass (*Panicum virgatum*), and prairie. The higher avian species richness in the prairie system (45 species; Fig. 11.2) might imply that prairies present a greater hazard to aircraft. However, when considering the relative hazard of the species found in the cover (Dolbeer et al. 2000, Dolbeer and Wright 2009, DeVault et al. 2011), corn had the greatest overall hazard to aviation (Fig. 11.2).

Federal Aviation Administration regulations dis-

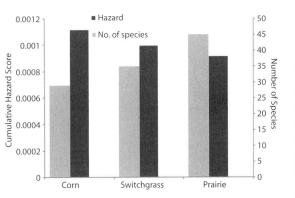

Fig. 11.2. Cumulative hazard and diversity of bird communities for three habitat types. Cumulative hazard scores were derived from relative hazard scores (Dolbeer and Wright 2009), were summed for each habitat type, and then scaled for interpretation. Lower values indicate less hazardous bird communities. Data adapted from Robertson et al. (2011)

courage the presence of "hazardous wildlife attractants," including all types of agriculture, at and near certificated U.S. airports (Federal Aviation Administration 2007, Blackwell et al. 2009). Even so, many U.S. airports lease portions of their land for agricultural production (Blackwell et al. 2009, DeVault et al. 2009), in part to reduce the economic burden of mowing turfgrass (Thomson 2007). These leased portions typically contain crops such as corn, wheat (*Triticum* spp.), and soybeans (*Glycine* spp.), which are wildlife attractants (Dolbeer et al. 1986, DeVault et al. 2007, Cerkal et al. 2009) even though they are notoriously depauperate, simplistic systems (Matson et al. 1997, Butler et al. 2007). If these systems lack diversity, then why are they not suited for airport use? These systems offer an important resource (i.e., food) for species that tend to be larger in size (e.g., white-tailed deer [*Odocoileus virginianus*]; Hein et al. 2012) and are most hazardous to aircraft (DeVault et al. 2011). But not all agriculture crops should be discounted categorically as a potential land cover for airports. Crops that lack palatable forage or abundant seed resources, such as some biofuel crops, may not attract hazardous wildlife, could potentially promote/protect some wildlife species of conservation concern, and provide some economic return. Empirical evidence is needed to determine which crops might fulfill these criteria at airports.

Herbaceous Cellulosic Feedstocks as a Potential Land Use at Airports

Crops under consideration for planting at airports include those that can be used to produce biofuel. Candidate crops for biofuel production range widely, from monocultures of exotic plants (e.g., *Miscanthus giganteus*; Heaton et al. 2008) to diverse native warm-season grass mixtures (Tilman et al. 2006, 2009; Somerville et al. 2010), although the use of nonherbaceous feedstocks may not be feasible within air operations areas (AOAs) because of safety concerns related to visibility (Austin-Smith and Lewis 1969). Existing grasslands at airports could potentially be managed for biofuel production if converted to appropriate herbaceous cellulosic feedstocks (Blackwell et al. 2009, DeVault et al. 2012). Switchgrass, for example, can yield 8.7–12.9 Mg/ha (19,180–28,440 lb/ha) of biomass depending on ecotype and management (McLaughlin and Kszos 2005, Adler et al. 2006, Mooney et al. 2008, Borsuk et al. 2010). Low-input, diverse native warm-season grass mixtures may produce even higher ethanol yields with greater greenhouse gas benefits than switchgrass monocultures (Tilman et al. 2006). The amount of grassland available at airports is much less than the area necessary to sustain a biofuel energy plant (Kocoloski et al. 2011), but airports could be integrated into an overall production and transportation strategy for biofuel production and thus could potentially contribute to this area of alternative energy production (DeVault et al. 2012).

Species composition of wildlife communities varies widely across different biofuel crops (Fargione et al. 2009, Meehan et al. 2010, Robertson et al. 2011). Field research is lacking on biofuel crops that, from an aviation perspective, would be compatible with safe airport operations, although research is ongoing (Blackwell et al. 2009, Martin et al. 2011, DeVault et al. 2012). We consider three possible land covers or grassland communities that might be feasible for the airport environment: switchgrass, *Miscanthus*, and a native prairie community (bluestems [*Andropogon* spp. and *Schizachyrium* spp.], Indiangrass [*Sorghastrum* spp.], and associated forbs).

Most research on herbaceous perennial grasslands for biofuels has been conducted on switchgrass (Murray and Best 2003, Murray et al. 2003, Roth et al. 2005;

Fig. 11.3. Switchgrass (*Panicum virgatum*) field planted for biomass production near West Point, Mississippi, USA. Photo credit: Tara Conkling

Fig. 11.3). But many of these studies were conducted on Conservation Reserve Program fields, which limit applicability to biofuel production at airports. Recent studies examining impacts of cellulosic biofuel crops on wildlife indicate that both *Miscanthus* and native grasses, including switchgrass and native warm-season grasses (as mentioned earlier), may provide benefits to some birds during winter and breeding seasons (Murray et al. 2003, Bellamy et al. 2009, Sage et al. 2010). The benefits of *Miscanthus* are temporary, however, without continuous wildlife management practices necessary to maintain the features of established plots that are attractive to birds (Bellamy et al. 2009). These features may be lost if plots are managed primarily to maximize biofuel production (Bellamy et al. 2009). There are additional questions regarding wildlife response to large plots of *Miscanthus* in the USA, as the vegetation structure is different from native grasslands, and it is unknown if avian species would perceive the bamboo-like vegetation as suitable habitat (Fargione et al. 2009).

Switchgrass and other native warm-season grasses may provide less ethanol output per unit area than *Miscanthus* (Heaton et al. 2008), but as native grass species, they might also be preferable as noninvasive wildlife habitat. Using switchgrass to convert existing row crop fields to biomass production provides new habitat for grassland birds (Murray et al. 2003), which could also reduce the presence of species typically attracted to crop fields (Dolbeer et al. 1986, DeVault et al. 2007). Roth et al. (2005) found that variation in the timing of switchgrass biofuel harvests and the resulting vegetation structure favored different grassland bird species, and a mosaic of harvest timings may increase local avian diversity. Recent research indicates that mixed-species grasslands with more diverse vegetation structures may provide even greater avian species richness and abundances than switchgrass (Robertson et al. 2011). T. J. Conkling et al. (unpublished data) have found prairie to be productive for breeding grassland birds such as dickcissels (*Spiza americana*), whereas switchgrass monoculture has demonstrated conservation value during winter months for species such as Le Conte's sparrow (*Ammodramus leconteii*). Preliminary results of studies in Mississippi investigating the hazard level of birds occupying switchgrass and prairie suggest these land covers may be suitable for airport grasslands in certain situations (T. J. Conkling et al., unpublished data).

Conservation of Birds

There are >3,300 km^2 (1,274 miles^2) of airport grasslands in the contiguous USA (DeVault et al. 2012).

Due to the amount of airport grasslands and because populations of grassland birds in North America are declining from habitat loss and degradation (Peterjohn and Sauer 1999, Askins et al. 2007), it has been suggested that airports may provide needed grassland habitat. However, airport grasslands pose challenges with respect to potential conservation efforts that must be recognized. We outline issues with habitat fragmentation, the role of airports as part of the general landscape, potential population losses of birds using airport grasslands, and the attraction of hazardous species to grasslands. Much of this section parallels the work of Blackwell et al. (2013).

Although the average airport in the contiguous USA contains 113 ha of turfgrass and other associated grassland cover types (DeVault et al. 2012), at many of these airports much of the grassland is scattered (i.e., fragmented) across a much larger area. Furthermore, some smaller airports do not contain grassland that extends appreciably beyond the AOA. The lack of large, unfragmented grassland tracts at some airports limits their value for grassland bird conservation. It is well established that habitat fragmentation negatively impacts abundance, distribution, and reproductive success of many grassland bird species, with declines more pronounced in area-sensitive species (Coppedge et al. 2001, Riffell et al. 2001, Chalfoun et al. 2002, Koper and Schmiegelow 2006, Ribic et al. 2009). Habitat fragmentation and the resulting loss of landscape connectivity is a major contributor to avian species declines and extinctions globally (Fischer and Lindenmayer 2007), yet patches as small as 50 ha may maximize bird species richness in a fragmented landscape (Helzer and Jelinski 1999), and small grassland patches with minimal edge habitat may also benefit grassland bird breeding and conservation (Davis and Brittingham 2004, Walk et al. 2010). Even so, research indicates that small grassland fragments cannot provide suitable habitat for bird species requiring large habitat patches (Johnson and Temple 1986, Vickery et al. 1995, Johnson and Igl 2001). Additionally, the shape of the habitat fragment and the distribution of fragments throughout the landscape can affect the settlement patterns of bird species (Laurance and Yensen 1991, Herkert 1994) or nest predation rates during the breeding season (Burger et al. 1994, Bergin et al. 2000, Grant et al. 2006). Therefore the habitat needs of the species

of interest must be compared to the available size and shape of grassland areas at each airport.

Local- and landscape-scale influences ultimately drive grassland bird use for most species (Cunningham and Johnson 2006, Blackwell et al. 2009, Martin et al. 2011). When considering the potential for airports as suitable habitat for grassland birds, airports must be viewed in association with the surrounding habitat matrix. In areas with substantial grassland surrounding patches, for example, nest success may increase (Berman 2007). Keyel et al. (2011) found that species believed to be area-sensitive may also respond to habitat openness, rather than patch size. If airports can provide additional grassland habitat to supplement the existing matrix, avian species—especially those with less stringent area requirements—may increase their use of these patches.

Despite the best intentions of biologists, conservation practices created specifically for wildlife on or off airport properties could result in sink habitats for grassland birds (McCoy et al. 1999, Murphy 2001). Ecological traps (Schlaepfer et al. 2002, Battin 2004) are also possible if infrequently managed grassland areas are mown during the breeding season (Kershner and Bollinger 1996), or if area-sensitive species are attracted to habitat patches with a high edge-to-area ratio (Winter and Faaborg 1999, Johnson and Igl 2001, Davis and Brittingham 2004, Renfrew et al. 2005). Some researchers argue that impacts to grassland species of conservation concern can be limited by adjusting timing of mowing relative to a species' breeding season (Brennan and Kuvlesky 2005). Kershner and Bollinger (1996) noted that nest predation accounted for only 23% of nest failures at airports in Illinois, relative to 44% of nest failures resulting from mowing. By altering mowing and providing some nest predator control, it may be possible to reduce the sink potential of airport grasslands for birds. Still, Blackwell et al. (2013) note that, regardless of whether airport grasslands function as sink habitats (Murphy 2001) or provide connectivity between grassland patches, issues associated with the attraction of species known to pose strike hazards to aviation remain (see also Martin et al. 2011).

Most grassland bird species require mature grasslands at some point in their life cycle (Askins et al. 2007); such habitats generally harbor greater invertebrate and vertebrate species diversity and richness

(Gardiner et al. 2002), which could also enhance resources for species hazardous to aviation (Sodhi 2002). Because safety should be the first priority of all airports, any grassland management approach that attracts hazardous species (DeVault et al. 2011) should be altered to reduce the attraction of the area to these species. If that alteration results in the loss of habitat for grassland bird species of concern, alternative management plans should be explored.

Grassland areas within the AOA may be minimally useful for grassland birds due to habitat fragmentation, small patch size, losses from mowing, and because providing permanent habitat for obligate grassland species will likely conflict with management techniques needed to remove food resources or roosting sites for hazardous species (Blackwell et al. 2013). One scenario that could possibly enhance grassland bird conservation, however, would be for grassland conservation management to occur beyond the AOA and other airport-specific siting criteria (Blackwell et al. 2009, 2013). Such placement might allow specific management of nonhazardous species on and near airport lands without compromising air safety.

Conservation of Mammals

Mammals are often overlooked as a source of risk for aviation, which has direct implications for conservation management of most mammalian species at airports. Dolbeer and Wright (2009) reported that, since 1990, U.S civil aircraft struck 36 mammal species, including eight species of bats. Of these 36 species, 21 (including two bat species; Dolbeer and Wright 2009) caused damage to aircraft. Mammal species considered high to extremely high hazards to aircraft included mule deer (*O. hemionus*), white-tailed deer, domestic dog (*Canis familiaris*), and coyote (*C. latrans*; Biondi et al. 2011, DeVault et al. 2011). Other mammal species struck by aircraft include eastern cottontail (*Sylvilgaris floridanus*), raccoon (*Procyon lotor*), black-tailed jackrabbit (*Lepus californicus*), woodchuck (*Marmota monax*), opossum (*Didelphus virginianus*), striped skunk (*Mephitis mephitis*), and red fox (*Vulpes vulpes*; K. M. Biondi, unpublished data; Dolbeer and Wright 2009). In addition to their high hazard ranking, the most frequently struck mammals are deer and coyotes (Dolbeer and Wright 2009, Biondi et al.

2011, DeVault et al. 2011). Any management or land-use modifications should avoid promoting use by deer and canids.

Mammal species of conservation concern are typically associated with unmanaged systems and are mostly ill adapted to human-altered environments (Ceballos et al. 2005), making mammal conservation at airports unlikely overall. Small mammals adapted for grasslands such as shrews (Soridae), *Peromyscus* spp., and other Muridae species—including cotton rats (*Sigmodon hispidus*) and jumping mice (*Zapus* spp.; Hall and Willig 1994, Kaufman et al. 1997)—may be attracted to airport grasslands. However, increased populations of these species at airports should generally be avoided, as both avian and mammalian predators of small mammals are typically large in size and hazardous to aircraft. Under simplistic models and assumptions, increased small-mammal diversity and biomass might cause functional and abundance shifts in predators (Holling 1965, Korpimäki and Norrdahl 1991, Korpimäki and Krebs 1996). Direct management of these predators may be possible, but the trade-off in conservation value, increased risk to aviation, and management cost would likely preclude targeted mammalian conservation at airports.

Summary

Conservation of wildlife species on airports, although problematic, may be best achieved through altering current land covers from traditional turfgrass management. Possible alternatives include prairie grass and switchgrass systems managed for forage or biofuels (DeVault et al. 2012). These options could, in some circumstances, conserve wildlife directly by providing in situ habitat for grassland birds (away from the AOA) or, perhaps more feasibly, indirectly by reducing the global carbon footprint (Tilman et al. 2009). Regardless, all alternative habitats at airports should be considered in the context of landscape fragmentation, metapopulation dynamics, and edge effects as they relate to grassland birds. Mammal conservation is not likely feasible at airports on any measurable scale. Most importantly, we encourage managers interested in wildlife conservation at airports to consider carefully how management of various grasslands systems might promote occupancy by hazardous species. Wildlife conservation will likely occur only past airport-specific siting criteria (Federal Aviation Administration 2007) to minimize risk to aviation (Blackwell et al. 2009, 2013). Potential economic benefits of alternative energy sources may contribute to adoption of biofuel grasslands on airports, but more research is needed.

LITERATURE CITED

Adler, P. R., M. A. Sanderson, A. A. Boateng, P. J. Weimer, and H. J. G. Jung. 2006. Biomass yield and biofuel quality of switchgrass harvested in fall or spring. Agronomy Journal 98:15–18.

Askins, R. A., F. Chavez-Ramirez, B. C. Dale, C. A. Haas, J. R. Herkert, F. L. Knopf, and P. D. Vickery. 2007. Conservation of grassland birds in North America: understanding ecological processes in different regions. Report of the AOU Committee on Conservation. Ornithological Monographs 64:1–46.

Austin-Smith, P. J., and H. F. Lewis. 1969. Alternative vegetative ground cover. Pages 153–160 in Proceedings of the world conference on bird hazards to aircraft, 2–5 September 1969. National Research Council of Canada, Kingston, Ontario, Canada.

Battin, J. 2004. When good animals love bad habitats: ecological traps and the conservation of animal populations. Conservation Biology 18:1482–1491.

Bellamy, P. E., P. J. Croxton, M. S. Heard, S. A. Hinsley, L. Hulmes, S. Hulmes, P. Nuttall, R. F. Pywell, and P. Rothery. 2009. The impact of growing Miscanthus for biomass on farmland bird populations. Biomass and Bioenergy 33:191–199.

Bergin, T. M., L. B. Best, K. E. Freemark, and K. J. Koehler. 2000. Effects of landscape structure on nest predation in roadsides of a midwestern agroecosystem: a multiscale analysis. Landscape Ecology 15:131–143.

Berman, G. M. 2007. Nesting success of grassland birds in fragmented and unfragmented landscapes of north central South Dakota. Thesis, South Dakota State University, Brookings, USA.

Biondi, K. M., J. L. Belant, J. A. Martin, T. L. DeVault, and G. Wang. 2011. White-tailed deer incidents with U.S. civil aircraft. Wildlife Society Bulletin 35:303–309.

Blackwell, B. F., T. L. DeVault, E. Fernández-Juricic, and R. A. Dolbeer. 2009. Wildlife collisions with aircraft: a missing component of land-use planning for airports. Landscape and Urban Planning 93:1–9.

Blackwell, B. F., T. W. Seamans, P. M. Schmidt, T. L. DeVault, J. L. Belant, M. J. Whittingham, J. A. Martin, and E. Fernández-Juricic. 2013. A framework for managing airport grasslands and birds amidst conflicting priorities. Ibis 155:199–203.

Blokpoel, H. 1976. Bird hazards to aircraft: problems and prevention of bird/aircraft collisions. Clarke, Irwin, Ottawa, Ontario, Canada.

Borsuk, M. E., S. D. Wullschleger, E. B. Davis, L. R. Lynd, and C. A. Gunderson. 2010. Biomass production in switchgrass across the United States: database description and determinants of yield. Agronomy Journal 102:1158–1168.

Brennan, L. A., and W. P. Kuvlesky. 2005. North American grassland birds: an unfolding conservation crisis? Journal of Wildlife Management 69:1–13.

Burger, L. D., J. Burger, and J. Faaborg. 1994. Effects of prairie fragmentation on predation on artificial nests. Journal of Wildlife Management 58:249–254.

Butler, S. J., J. A. Vickery, and K. Norris. 2007. Farmland biodiversity and the footprint of agriculture. Science 315:381–384.

Ceballos, G., P. R. Ehrlich, J. Soberón, I. Salazar, and J. P. Fay. 2005. Global mammal conservation: what must we manage? Science 309:603–607.

Cerkal, R., J. Vejrazka, J. Kamler, and J. Dvorak. 2009. Game browse and its impact on selected grain crops. Plant Soil Environment 55:181–186.

Chalfoun, A. D., F. R. Thompson III, and M. J. Ratnaswamy. 2002. Nest predators and fragmentation: a review and meta-analysis. Conservation Biology 16:306–318.

Conover, M. R. 2002. Resolving human–wildlife conflicts: the science of wildlife damage management. CRC Press, Boca Raton, Florida, USA.

Coppedge, B. R., D. M. Engle, R. E. Masters, and M. S. Gregory. 2001. Avian response to landscape change in fragmented southern Great Plains grasslands. Ecological Applications 11:47–59.

Cunningham, M. A., and D. H. Johnson. 2006. Proximate and landscape factors influence grassland bird distributions. Ecological Applications 16:1062–1075.

Davis, S. K., and M. Brittingham. 2004. Area sensitivity in grassland passerines: effects of patch size, patch shape, and vegetation structure on bird abundance and occurrence in southern Saskatchewan. Auk 121:1130–1145.

DeVault, T. L., J. C. Beasley, L. A. Humberg, B. J. MacGowan, M. I. Retamosa, and O. E. Rhodes Jr. 2007. Intrafield patterns of wildlife damage to corn and soybeans in northern Indiana. Human–Wildlife Conflicts 1:205–213.

DeVault, T. L., J. L. Belant, B. F. Blackwell, J. A. Martin, J. A. Schmidt, L. W. Burger, and J. W. Patterson. 2012. Airports offer unrealized potential for alternative energy production. Environmental Management 49:517–522.

DeVault, T. L., J. L. Belant, B. F. Blackwell, and T. W. Seamans. 2011. Interspecific variation in wildlife hazards to aircraft: implications for airport wildlife management. Wildlife Society Bulletin 35:394–402.

DeVault, T. L., J. E. Kubel, O. E. Rhodes Jr., and R. A. Dolbeer. 2009. Habitat and bird communities at small airports in the midwestern USA. Proceedings of the Wildlife Damage Management Conference 13:137–145.

Dolbeer, R. A., P. P. Woronecki, and R. L. Bruggers. 1986. Reflecting tapes repel blackbirds from millet, sunflowers, and sweet corn. Wildlife Society Bulletin 14:418–425.

Dolbeer, R. A., and S. E. Wright. 2009. Safety management systems: how useful will the FAA National Wildlife Strike Database be? Human–Wildlife Conflicts 3:167–178.

Dolbeer, R. A., S. E. Wright, and E. C. Cleary. 2000. Ranking the hazard level of wildlife species to aviation. Wildlife Society Bulletin 28:372–378.

Fahrig, L. 2003. Effects of habitat fragmentation on biodiversity. Annual Review of Ecology, Evolution, and Systematics 34:487–515.

Fargione, J. E., T. R. Cooper, D. J. Flaspohler, J. Hill, C. Lehman, T. McCoy, S. McLeod, E. J. Nelson, K. S. Oberhauser, and D. Tilman. 2009. Bioenergy and wildlife: threats and opportunities for grassland conservation. Bioscience 59:767–777.

Federal Aviation Administration. 2007. Hazardous wildlife attractants on or near airports. Advisory Circular 150/5200-33B. U.S. Department of Transportation, Washington, D.C., USA.

Fischer, J., and D. B. Lindenmayer. 2007. Landscape modification and habitat fragmentation: a synthesis. Global Ecology and Biogeography 16:265–280.

Gardiner, T., M. Pye, R. Field, and J. Hill. 2002. The influence of sward height and vegetation composition in determining the habitat preferences of three Chorthippus species (Orthoptera: Acrididae) in Chelmsford, Essex, UK. Journal of Orthoptera Research 11:207–213.

Grant, T. A., E. M. Madden, T. L. Shaffer, P. J. Pietz, G. B. Berkey, and N. J. Kadrmas. 2006. Nest survival of clay-colored and vesper sparrows in relation to woodland edge in mixed-grass prairies. Journal of Wildlife Management 70:691–701.

Groom, M. J., G. K. Meffe, and C. R. Carroll. 2006. Principles of conservation biology. Sinauer, Sunderland, Massachusetts, USA.

Hall, D. L., and M. R. Willig. 1994. Mammalian species composition, diversity, and succession in conservation reserve program grasslands. Southwestern Naturalist 39:1–10.

Hall, L. S., P. R. Krausman, and M. L. Morrison. 1997. The habitat concept and a plea for standard terminology. Wildlife Society Bulletin 25:173–182.

Heaton, E. A., F. G. Dohleman, and S. P. Long. 2008. Meeting US biofuel goals with less land: the potential of *Miscanthus*. Global Change Biology 14:2000–2014.

Hein, A. M., C. Hou, and J. F. Gillooly. 2012. Energetic and biomechanical constraints on animal migration distance. Ecology Letters 15:104–110.

Helzer, C. J., and D. E. Jelinski. 1999. The relative importance of patch area and perimeter-area ratio to grassland breeding birds. Ecological Applications 9:1448–1458.

Herkert, J. R. 1994. The effects of habitat fragmentation on midwestern grassland bird communities. Ecological Applications 4:461–471.

Holling, C. S. 1965. The functional response of predators to prey density and its role in mimicry and population regulation. Memoirs of the Entomological Society of Canada 97:1–60.

International Civil Aviation Organization. 2002. Land use and environmental control. Airport Planning Manual 9184 AN/902, Part 2. Montreal, Quebec, Canada.

Johnson, D. H., and L. D. Igl. 2001. Area requirements for grasslands birds: a regional perspective. Auk 118:24–34.

Johnson, R. G., and S. A. Temple. 1986. Assessing habitat quality for birds nesting in fragmented tallgrass prairies. Pages 245–249 in J. Verner, M. L. Morrison, and C. J. Ralph, editors. Wildlife 2000: modeling habitat relationships of terrestrial vertebrates. University of Wisconsin Press, Madison, USA.

Kaufman, G. A., D. W. Kaufman, F. L. Knopf, and F. B. Samson. 1997. Ecology of small mammals in prairie landscapes. Pages 207–243 in F. L. Knopf and F. B. Samson, editors. Ecology and conservation of Great Plains vertebrates. Springer, New York, New York, USA.

Kelly, T., and J. Allan. 2006. Ecological effects of aviation. The ecology of transportation: managing mobility for the environment. Environmental Pollution 10:5–24.

Kershner, E. L., and E. K. Bollinger. 1996. Reproductive success of grassland birds at east-central Illinois airports. American Midland Naturalist 136:358–366.

Kessel, B. 1957. A study of the breeding biology of the European starling (*Sturnus vulgaris* L.) in North America. American Midland Naturalist 58:257–331.

Keyel, A. C., C. M. Bauer, C. R. Lattin, L. M. Romero, and J. M. Reed. 2011. Testing the role of patch openness as a causal mechanism for apparent area sensitivity in a grassland specialist. Oecologia 169:1–12.

Kocoloski, M., W. M. Griffin, and H. S. Matthews. 2011. Impacts of facility size and location decisions on ethanol production cost. Energy Policy 39:47–56.

Koper, N., and F. K. K. Schmiegelow. 2006. A multi-scaled analysis of avian response to habitat amount and fragmentation in the Canadian dry mixed-grass prairie. Landscape Ecology 21:1045–1059.

Korpimäki, E., and C. J. Krebs. 1996. Predation and population cycles of small mammals. BioScience 46:754–764.

Korpimäki, E., and K. Norrdahl. 1991. Numerical and functional responses of kestrels, short-eared owls, and long-eared owls to vole densities. Ecology 72:814–826.

Laurance, W. F., and E. Yensen. 1991. Predicting the impacts of edge effects in fragmented habitats. Biological Conservation 55:77–92.

Martin, J. A., J. L. Belant, T. L. DeVault, B. F. Blackwell, L. W. Burger Jr., S. K. Riffell, and G. Wang. 2011. Wildlife risk to aviation: a multi-scale issue requires a multi-scale solution. Human–Wildlife Interactions 5:198–203.

Matson, P. A., W. J. Parton, A. G. Power, and M. J. Swift. 1997. Agricultural intensification and ecosystem properties. Science 277:504–509.

McCoy, T. D., M. R. Ryan, E. W. Kurzejeski, and L. W. Burger Jr. 1999. Conservation Reserve Program: source or sink habitat for grassland birds in Missouri? Journal of Wildlife Management 63:530–538.

McLaughlin, S. B., and L. A. Kszos. 2005. Development of switchgrass (*Panicum virgatum*) as a bioenergy feedstock in the United States. Biomass and Bioenergy 28:515–535.

Meehan, T. D., A. H. Hurlbert, and C. Gratton. 2010. Bird communities in future bioenergy landscapes of the Upper Midwest. Proceedings of the National Academy of Sciences of the United States of America 107:18,533–18,538.

Meyer, W. B. 1995. Past and present land use and land cover in the USA. Consequences 1:25–33.

Mooney, D. F., R. K. Roberts, B. C. English, D. D. Tyler, and J. A. Larson. 2008. Switchgrass production in marginal environments: a comparative economic analysis across four west Tennessee landscapes. Proceedings of the 2008 annual meeting. Agricultural and Applied Economics Association, 27–29 July 2008, Orlando, Florida, USA.

Murphy, M. T. 2001. Source–sink dynamics of a declining eastern kingbird population and the value of sink habitats. Conservation Biology 15:737–748.

Murray, L. D., and L. B. Best. 2003. Short-term bird response to harvesting switchgrass for biomass in Iowa. Journal of Wildlife Management 67:611–621.

Murray, L. D., L. B. Best, T. J. Jacobsen, and M. L. Braster. 2003. Potential effects on grassland birds of converting marginal cropland to switchgrass biomass production. Biomass and Bioenergy 25:167–175.

Peterjohn, B. G., and J. R. Sauer. 1999. Population status of North American grassland birds from the North American breeding bird survey, 1966–1996. Studies in Avian Biology 19:27–44.

Renfrew, R. B., C. A. Ribic, and J. L. Nack. 2005. Edge avoidance by nesting grassland birds: a futile strategy in a fragmented landscape. Auk 122:618–636.

Ribic, C. A., R. R. Koford, J. R. Herkert, D. H. Johnson, N. D. Niemuth, D. E. Naugle, K. K. Bakker, D. W. Sample, and R. B. Renfrew. 2009. Area sensitivity in North American grassland birds: patterns and processes. Auk 126:233–244.

Riffell, S. K., B. E. Keas, and T. M. Burton. 2001. Area and habitat relationships of birds in Great Lakes coastal wet meadows. Wetlands 21:492–507.

Robertson, B. A., P. J. Doran, L. R. Loomis, J. R. Robertson, and D. W. Schemske. 2011. Perennial biomass feedstocks enhance avian diversity. GCB Bioenergy 3:235–246.

Roseberry, J. L., and W. D. Klimstra. 1970. The nesting ecology and reproductive performance of the Eastern Meadowlark. Wilson Bulletin 82:243–267.

Roth, A. M., D. W. Sample, C. A. Ribic, L. Paine, D. J. Undersander, and G. A. Bartelt. 2005. Grassland bird response to harvesting switchgrass as a biomass energy crop. Biomass and Bioenergy 28:490–498.

Sage, R., M. Cunningham, A. J. Haughton, M. D. Mallot, D. A. Bohan, A. Riche, and A. Karp. 2010. The environmental impacts of biomass crops: use by birds of *Miscanthus* in summer and winter in southwestern England. Ibis 152:487–499.

Schlaepfer, M. A., M. C. Runge, and P. W. Sherman. 2002. Ecological and evolutionary traps. Trends in Ecology and Evolution 17:474–480.

Sodhi, N. S. 2002. Competition in the air: birds versus aircraft. Auk 119:587–595.

Somerville, C., H. Youngs, C. Taylor, S. C. Davis, and S. P. Long. 2010. Feedstocks for lignocellulosic biofuels. Science 329:790–792.

Thomson, B. 2007. A cost effective grassland management strategy to reduce the number of bird strikes at the Brisbane airport. Thesis, Queensland University of Technology, Brisbane, Australia.

Tilman, D., J. Hill, and C. Lehman. 2006. Carbon-negative biofuels from low-input high-diversity grassland biomass. Science 314:1598–1600.

Tilman, D., R. Socolow, J. A. Foley, J. Hill, E. Larson, L. Lynd, S. Pacala, J. Reilly, T. Searchinger, C. Somerville, and R. Williams. 2009. Beneficial biofuels—the food, energy, and environment trilemma. Science 325:270–271.

Turner, M. G., R. H. Gardner, and R. V. O'Neill. 2001. Landscape ecology in theory and practice: pattern and process. Springer-Verlag, New York, New York, USA.

Vickery, P. C., M. L. Hunter, and S. M. Melvin. 1995. Effects of habitat area on the distribution of grassland birds in Maine. Conservation Biology 8:1087–1097.

Walk, J. W., E. L. Kershner, T. J. Benson, and R. E. Warner. 2010. Nesting success of grassland birds in small patches in an agricultural landscape. Auk 127:328–334.

Winter, M., and J. Faaborg. 1999. Patterns of area sensitivity in grassland-nesting birds. Conservation Biology 13:1424–1436.

PART III · WILDLIFE MONITORING

12

Jerrold L. Belant
Brian E. Washburn
Travis L. DeVault

Understanding Animal Movements at and near Airports

Understanding movements of hazardous wildlife species at and near airports is critical to formulating effective management strategies for reducing aviation risk. Animal movements vary daily, seasonally, and annually and are based on broad biological and ecological concepts, including foraging, reproduction, habitat characteristics, dispersal, and migration. As an energy conservation strategy, most animals minimize their movements to meet life requisites, which in turn presumably improves fitness. Animal movements in relation to airports can be direct; for example, Canada geese (*Branta canadensis*) flying onto an airfield because grass height and composition are suitable for loafing sites and as food. Animal movements in and around airports can also be indirect; for example, airports near large rivers may experience increased numbers of birds flying overhead during spring and autumn migrations, as rivers often facilitate bird navigation.

In this chapter we describe ecologically based patterns of animal movements and develop a mechanistic foundation for understanding those movements and the degree to which we can modify them to reduce corresponding hazards to aircraft. We discuss biological and ecological causes of animal movements and some of the foundational ecological theories that help explain animal movements at airports. We then discuss motivations of animal movements at airports based on resource needs, the role of spatial scale when considering animal movements, and how to apply these concepts to reduce wildlife strikes. We end with a brief description of primary techniques to quantify animal movements, summarize management of animal movements at airports, and suggest areas of future research.

Types of Animal Movements

Animal movements can be divided into six broad, ecologically based categories: foraging, movements to rest sites, reproduction, territory defense, dispersal, and migration. We generally define foraging as any animal movement to feed, to obtain free water for drinking, or to search for food. Movements to rest sites are those where animals are seeking shelter (e.g., night roosts for turkey vultures [*Cathartes aura*] or bedding sites for white-tailed deer [*Odocoileus virginianus*]). Reproduction movements are associated with individuals searching for mates during a defined breeding season (e.g., white-tailed deer during the rut). Defense movements are those in which an animal is defending either a territory or a specific resource (e.g., food) from conspecifics or other animals. Dispersal includes movements of juvenile individuals traveling from their natal range to locate new areas to occupy (Greenwood 1980, Waser and Jones 1983, Clutton-Brock 1989, Waser 1996). Migrations are typically biannual movements of animals in response to changes in resource availability and for reproduction; for example, the spring and fall migrations of many bird species (Drent et al. 2003, van Wijk et al. 2012). These categories of movement vary temporally and spatially. Foraging occurs at least daily for most species, whereas migration typically oc-

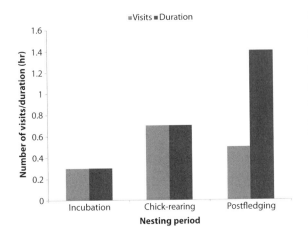

Fig. 12.1. Number and duration of visits to a landfill in northern Ohio by nesting period of radio-tagged, nesting herring gulls (*Larus argentatus*). Derived from Belant et al. (1993)

curs twice annually, and dispersal by definition occurs once in a lifetime. In a spatial context, movements for foraging tend to be more restricted than movements to rest sites (but not always), which in turn are more restricted than defense and reproduction, dispersal, and migration. These categories can also be hierarchical; for example, foraging tends to occur during reproduction, dispersal, and migration.

During the nesting and young-rearing periods, most adult birds and many mammals behave as central place foragers (Orians and Pearson 1979, Kacelnik 1984, Olsson et al. 2008, Wakefield et al. 2009), in that they return repeatedly to the nest or den site to provision young with food obtained during foraging bouts. For birds, these movements can vary in frequency and duration among incubation, chick-rearing, and postfledging periods. The mean daily number of visits to a landfill by radio-tagged, nesting herring gulls (*Larus argentatus*) generally increased in frequency and duration from incubation to postfledging periods—a consequence of energy demands of the chicks and reduced tenacity to the nest site after the young fledged (Belant et al. 1993; Fig. 12.1). These movements can in turn influence use of airports, either directly through increased foraging bouts during chick rearing or indirectly as birds fly over the airport to seek resources.

Wildlife managers must consider that how, when, and where animal movements occur are based fundamentally in natural selection. Animals use resources (e.g., food, rest sites, mates) to help ensure their sur-

vival; greater survival will often result in greater recruitment of young, which is how species persist. Success is based on how resources necessary for survival are distributed across the landscape, and how well animals adapt to changing distributions. Animals in part reduce energetic costs by minimizing movements required to acquire these necessary resources, which can be a better predictor of fitness than traditional habitat selection (e.g., Ayers et al. 2013). Recently developed spatial energetic models can assess landscapes relative to a species' resource needs to better understand species' movements and distributions (e.g., Wilson et al. 2012). Spatial energetic models applied to landscapes have particular application to the evaluation of management scenarios that might reduce resources and, subsequently, wildlife risk to aircraft. Understanding species movements and distributions could be used to refine habitat management practices to reduce animal use in and around airports.

Principles of Animal Movements

There are numerous ecological theories and processes that relate to spatial and temporal aspects of animal movements. Several of the more fundamental theories have strong application to animal movements in relation to management at airports. Understanding these principles will help airport biologists and managers develop and implement strategies to reduce animal movements at airports. We provide basic definitions and demonstrate their application to airport management.

Distribution Theory

Animal distributions are grounded within two pervasive models: ideal free distribution and ideal despotic distribution. The ideal free distribution model generally applies to nonterritorial animals and states that individuals are distributed proportionately to resources (e.g., roost sites, foraging sites) available (Fretwell and Lucas 1970). In this model, animals assess the quality of available resources and move unhindered among these resources to select those considered best. In contrast, the ideal despotic distribution model applies to territorial animals whereby dominant individuals influence amount of resources available to subordinate individuals (Fretwell 1972). Subordinates' selection of habitat is consequently constrained by the aggressive behavior and distribution of dominant animals. A practical goal of wildlife management at airports would be to follow one or both of these models to evaluate and reduce wildlife use, especially use by hazardous wildlife species. To limit movements of wildlife at airports under the ideal free distribution model, a reduction in suitable resources would be necessary. This could involve reduction or removal of food sources (e.g., Bernhardt et al. 2009, Washburn et al. 2011; Chapter 8) or roosting areas (e.g., Gordon and White 2006). In these situations, animals will seek areas other than the airport to obtain food or to locate another roost. More direct management actions (e.g., harassment, exclusion) would follow the ideal despotic distribution model, whereby humans would be the dominant individuals (i.e., despots) and constrain use of airport resources by hazardous wildlife (subordinates). This would be accomplished through aggressive behavior in the case of harassment techniques (e.g., Montoney and Boggs 1993), or through human presence (i.e., distribution) in the case of fencing to exclude wildlife (e.g., DeVault et al. 2008; Chapter 5). The relevant principle for both models is to reduce wildlife movements at airports by either reducing resource quality or constraining wildlife movements through management actions.

Niche Theory

As with distribution theory, niche theory has considerable application to animal movements at and near airports. It describes the role of an organism in its environment (e.g., predator, parasite), including its activities and interrelationships with other organisms (Krebs 2001). The set of resources that a species can use in the absence of competition or other interactions with animals has been termed the fundamental niche (Krebs 2001). Because of interactions with other animals, however, individuals and species typically are restricted to a narrower range of ecological or resource conditions. This restricted range of conditions is referred to as the realized niche (Caughley and Sinclair 1994). Essentially all animals operate within a realized niche, being constrained by competition with other animals, environmental limitations, and other factors. Finally, the range of resource conditions (e.g., number of available resting sites, amount of food available) that an animal can use and still persist in the environment has been coined the niche hypervolume (Hutchinson 1957). Reduction of animal movements (or use) in airport environments will require a great enough reduction in one or more of the resources at the airport that is within an animal's realized niche, such that the animal will no longer access the airport to search for these resources. If multiple resources (e.g., food, shelter) are available at the airport, animal movements onto airport property may be reduced only after all suitable resources are adequately managed.

Foraging Theory

It has been suggested frequently that animals optimize their foraging activities to increase their odds of survival (Schoener 1971, Krebs 1973). A part of optimal foraging is the marginal value theorem (Charnov 1976), which in its most fundamental form states that an animal will occupy a suitable area of habitat until resource depletion (i.e., to a particular resource density) by that individual causes it to move to another area of higher habitat quality. This response by animals has been referred to as the "giving-up density." An important point of this theorem is that the giving-up density of an area occupied by an animal will depend in part on the distance to the next suitable area. An animal is more likely to stay in the current habitat longer if the next area of suitable habitat is farther away, which has implications for wildlife harassment (methods that can increase perceived risk and therefore the giving-up density; see Brown 1999) at airports. If another suitable area is a considerable distance from the airport, animals will be less likely to disperse from the area or will be more likely to return.

Effects of Group Size

Animals congregate in groups for numerous reasons: to rear young, to reduce risk of predation, and to procure food (Heinsohn 1991, Sirot and Touzalin 2009, Thornton and Clutton-Brock 2011). Whether animals move as individuals or in groups is of great importance to airport managers, as the likelihood of aircraft damage generally increases with the number of animals struck. Biondi et al. (2011) reported that aircraft were 25 times

Fig. 12.2. European starlings (*Sturnus vulgaris*) are a moderate risk to aircraft, but the likelihood of damage increases markedly when starlings form large flocks. Starlings often flock during foraging and roosting. Photo credit: Tommy Hansen

more likely to be damaged during incidents involving multiple white-tailed deer compared to strikes with a single animal. Although individual European starlings (*Sturnus vulgaris*) are considered a moderate hazard to aircraft relative to other wildlife species, with 4% of strikes causing damage, a high proportion of damaging strikes was a consequence of aircraft colliding with multiple individuals (Dolbeer and Wright 2009; Fig. 12.2). Notably, multiple Canada geese resulted in the forced landing of US Airways Flight 1549 on 15 January 2009 (Marra et al. 2009). The primary causes for increased damage to aircraft from hitting multiple animals appear to be related to species body mass (e.g., DeVault et al. 2011) and multiple strike locations on the aircraft.

Motivations for Animal Movements at and near Airports

Motivations for animal movements at airports can be characterized into three broad categories. The first is movement in response to habitat or other features that may cause attraction (e.g., foraging or roosting site) or avoidance (e.g., avoiding aircraft or buildings) of airports. Laughing gulls (*L. atricilla*) nesting at Jamaica Bay National Wildlife Refuge apparently make daily foraging trips from the nesting colony to loaf or forage on beetles and ants at and near John F. Kennedy In-

ternational Airport (JFK; Buckley and McCarthy 1994, Bernhardt et al. 2010, Kutschbach-Brohl et al. 2010). In this situation, foraging movements to obtain terrestrial invertebrates varied during summer, with greatest apparent movements during July (Bernhardt et al. 2010), presumably when adults were provisioning young (Dolbeer et al. 1993, Washburn et al. 2013). Tree swallow (*Tachycineta bicolor*) use of northern bayberry fruit during autumn at JFK resulted in extensive use of this resource by large flocks of swallows, causing a seasonal hazard to aircraft (Bernhardt et al. 2009).

The second category includes movements at or adjacent to airports that may be completely unrelated to the airport, including bird migrations or daily flights from roosting to foraging sites. Servoss et al. (2000) documented large flocks of blackbirds (Icteridae) and European starlings flying over Phoenix Sky Harbor International Airport to reach attractive habitats outside the airport boundaries. Nohara et al. (2011) documented with radar flocks of Canada geese and other bird species crossing airspace at JFK. Movements of these types can be more difficult to manage, as the cause of animal movements is not necessarily a consequence of habitat or other resources on the airport; rather, bird movements across airports are artifacts of the airport location in relation to other landscape features.

The third category of animal movements is a response to direct or indirect wildlife control actions (e.g., hazing birds from runways, white-tailed deer movements along perimeter fences). In these cases, wildlife movement can be considered constrained (or modified) from movements that would ordinarily occur without management. For example, suspending vulture effigies from roosts reduced vulture use of U.S. Marine Corps Air Station, Cherry Point, North Carolina, USA (Ball 2009). Each of these three movement categories varies markedly in terms of effective management techniques and strategies to reduce risk to aircraft. Consequently, understanding the causes of animal movements at airports is critical for development of appropriate management strategies.

Integrating Spatial Scale

Most of the early wildlife management techniques to reduce wildlife strikes with aircraft occurred only on airport properties. These approaches included both harassment and habitat management techniques (e.g., maintaining a specific grass height; International Civil Aviation Organization 1991; Chapters 8 and 10). Yet few wildlife species of high risk to aviation spend all of their time on airport property. Consequently, greater emphasis has been placed in recent years on management of areas surrounding airports (see Blackwell et al. 2009, Dolbeer 2011). Martin et al. (2011) highlighted the importance of spatial scale relative to animal movements and types of movements (e.g., feeding, migration). Animal movements in relation to airports can be considered in a hierarchical structure that includes multiple spatial extents (Martin et al. 2011); these spatial extents should correspond to types of movements (e.g., foraging, dispersal) for each species considered hazardous to aircraft. Davis et al. (2003) developed a risk-based model in an effort to establish zoning criteria for land use near Canadian airports. These authors suggested a framework that considered existing land-use practices, bird species characteristics linked to aircraft safety (e.g., body size, flocking behavior), and relative risk of aircraft during varying phases of flight (see also Blackwell et al. 2009). In principle, this framework would reduce suitability of habitats near airports and consequently reduce use (i.e., movements) of animals hazardous to aircraft in these areas. Others have recognized this basic premise; for example, the Federal Aviation Administration (FAA) currently provides separation criteria for hazardous wildlife attractions (e.g., landfills) at or near airports with a maximum distance of 8 km (5 miles; Dolbeer 2006, FAA 2007). However, these guidelines do not take into account species-specific movements relative to foraging or other behaviors. Belant et al. (1993, 1998) documented herring gull and ring-billed gull (*L. delawarensis*) movements up to 26 km (16 miles) from the nesting colony to landfills to acquire food. York et al. (2000*a*,*b*) similarly determined that Canada geese in Alaska sometimes moved distances >15 km (9 miles) from molting sites to airports to loaf and forage. Although highly variable, the number of marked geese observed at Elmendorf Air Force Base declined as distance from the original molting site increased (Fig. 12.3). In all of these studies, relative use of sites (landfills or airports) decreased as distance increased from source locations (nesting colony and molting sites). Nevertheless, animal movements to acquire food or secure loafing sites

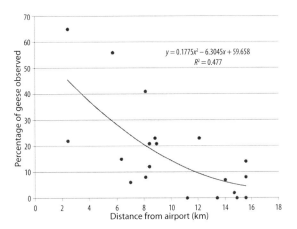

Fig. 12.3. Percentage of Canada geese (*Branta canadensis*) observed at Elmendorf Air Force Base, Anchorage, Alaska, during summer in relation to distance from the original capture site. Derived from York et al. (2000b)

were about twice the maximum distance specified by the FAA for hazardous wildlife attractions.

An important consideration is that larger species within a taxonomic group (e.g., birds, mammals) generally have greater local movements (e.g., when foraging) as well as dispersal and migration movements compared to smaller species (Harestad and Bunnell 1979, Mace and Harvey 1983, Lindstedt et al. 1986, Basset 1995, Silva and Downing 1995, Kelt and Van Vuren 1999, Hein et al. 2012). Efforts to reduce risk to aircraft must occur at a spatial scale much larger than the airport and must consider distances moved by hazardous wildlife species. They will also require landscape-level planning that integrates information on species' movements and habitat needs from ecologists, from airport managers relative to hazardous wildlife species, and from other private and government entities relative to potential land management practices (Belant 1997, Blackwell et al. 2009).

Applications for Reducing Wildlife Strikes

Understanding the types and causes of movement can improve our ability to manage wildlife at and near airports, which in turn can reduce risk to aircraft. Bernhardt et al. (2009) conducted an excellent example of incorporating a mechanistic understanding of animal movements to reduce hazards to aircraft at JFK. Tree swallows were involved in 109 strikes with U.S. civil aircraft in airport environments from 1990 to 2009 (DeVault et al. 2011). Although their relative hazard score to aircraft is low (Dolbeer and Wright 2009, DeVault et al. 2011), large flocks of tree swallows represent a hazard to aircraft at JFK, especially during autumn (Dolbeer et al. 2003). To address this issue, Bernhardt et al. (2009) determined that the diet of tree swallows during autumn was predominantly northern bayberry fruit (*Myrica pensylvanica*). The airport initiated a bayberry removal program, removing 75% of bushes within 0.8 km (0.5 miles) of the runway and about 50% of bushes elsewhere on the airport. During the seven years following initiation of bayberry removal, aircraft collisions with tree swallows were reduced by 75% (Bernhardt et al. 2009). The reduction in bayberry bushes reduced food availability for tree swallows, which in turn reduced swallow movements at the airport.

An example of reducing wildlife risk to aircraft, where birds crossed the airfield to forage and loaf at sites beyond airport property, involved gulls (particularly laughing gulls) at JFK (Dolbeer et al. 1989, 1993). Gulls were involved in 87% (laughing gulls 52%) of aircraft strikes at JFK from 1988 to 1990 (Dolbeer et al. 1993), with most strikes occurring during May–September and peaking during June–July, when laughing gulls were nesting (Washburn et al. 2012). An integrated gull-strike reduction program with a lethal control component (i.e., shooting program) has been implemented at JFK since 1991; this program reduced the number of laughing gull–aircraft collisions by 62% in 1991 and 76–99% annually from 1992 to 2008, compared with the mean of 157 strikes per year from 1988 to 1990 (Dolbeer et al. 1993, Washburn et al. 2009; Chapter 7). Attempts to change gull movement patterns by reducing suitability of foraging and loafing sites was considered untenable, because laughing gulls access these sites throughout the metropolitan New York City area (Griffin and Hoopes 1991, Washburn et al. 2013). Of interest is that movements of laughing gulls did not suggest avoidance of JFK during the first years of lethal control (Dolbeer et al. 1993); however, gulls in later years of the control program altered their flight patterns in response to control efforts (Dolbeer et al. 2003). Gulls apparently recognized shooters as a risk, as evidenced by their avoidance of people standing with guns at the airport boundary but not shooting (Barras et al. 2000). Avoidance of animals due to predation risk is a learned behavior (Sirot 2010) that allows species to rapidly identify predators and to invoke antipredator strategies (Lonnstedt et al. 2012), in this case by avoiding shooters on the airport (see also Chapters 2 and 4).

Techniques for Investigating Animal Movements at Airports

Numerous techniques are available to estimate and model animal abundance and distributions that can be applied to airport environments; however, far fewer techniques are available to estimate animal movements. Most wildlife survey and monitoring techniques emphasize one or more elements of species occurrence (e.g., MacKenzie et al. 2006; Chapter 14), from which animal movement can be inferred but not directly measured. The two primary techniques to study animal movements involve direct use of radiotelemetry (Millspaugh et al. 2012) and radar (Chapter 13). Radiotelemetry can provide finer spatial resolution and is based on information obtained from individual animals. In contrast, radar often results in slightly coarser spatial resolution of animal movements and species identification (Beason et al. 2010). The best technique will depend on the specific goals and objectives for each airport.

Wildlife radiotelemetry has been one of the most effective techniques in understanding animal ecology, including information on animal locations and movements. The types of radiotelemetry most applicable to understanding animal movements at airports include very high frequency (VHF) transmitters and satellite telemetry platforms. For VHF systems, transmitters are attached to animals and emit a unique radio frequency that personnel can locate manually by using a specialized receiver. Satellite-telemetry units rely on a constellation of satellites to obtain animal locations and offer the ability to estimate locations of animals on the ground and in the air (Tomkiewicz et al. 2010, Washburn and Olexa 2011). Radiotelemetry has been used to estimate animal movements in relation to aviation risk on several occasions. Schafer et al. (2002) used VHF and ARGOS satellite radiotelemetry to estimate the effectiveness of translocating red-tailed hawks (*Buteo jamaicensis*) from Chicago O'Hare International

Airport, Chicago, Illinois, USA. Similarly, Schumacher et al. (2008) estimated movements of translocated immature bald eagles (*Haliaeetus leucocephalus*) in relation to aviation risk using Global Positioning System (GPS) satellite telemetry. York et al. (2000*b*) estimated movements of Canada geese between molt sites and an Alaskan airport using VHF radiotelemetry. Washburn and Olexa (2011) used information from GPS satellite telemetry units attached to ospreys (*Pandion haliaetus*) to develop three-dimensional airspace risk models and to quantify the risk of osprey collisions with military aircraft during both breeding and migratory seasons.

In some situations, radar can be used to estimate timing, trajectories, flock size, altitudes, and speeds traveled (Klope et al. 2009, Nohara et al. 2011; Chapter 13). Radar has been used to estimate the distribution of birds over airfields as well as the frequency of near misses between birds and aircraft (MacKinnon 2006, Klope et al. 2009, FAA 2010; Chapter 13). A potential advantage of using radar at airports is a move toward near-real-time detection of birds, which could help alert airport biologists of developing threats (Blokpoel and MacKinnon 2011, Nohara et al. 2011; Chapter 13).

Managing Animal Movements at Airports

Effective management of hazardous wildlife at airports requires sound information on species presence and abundance (or relative abundance) in relation to the relative hazard each species represents to aircraft (e.g., Dolbeer et al. 2000, 2010; Biondi et al. 2011; DeVault et al. 2011; Chapter 14). In addition, detailed information on actual wildlife strikes at individual airports is necessary. Once this information is obtained, it can help managers understand the ecological reasons (e.g., to forage) hazardous species use airport property. It is typically recommended that airports direct management efforts toward the species most hazardous to aircraft (Dolbeer and Wright 2009).

Managing wildlife at airports generally involves use of indirect (e.g., fences) or direct (e.g., harassment) actions to discourage animal use of resources. Indirect management techniques include reducing food availability (Chapter 8) or the presence of water (Chapter 9), manipulating existing vegetation (e.g., turfgrass; Chapter 10), and using exclusion devices (Chapter 5). Direct management actions include the use of visual, chemical, tactile,

or auditory deterrents (Chapters 2–4) and translocation of hazardous wildlife (Chapter 6). Using multiple methods often maximizes the effectiveness of wildlife control techniques (Conover 2002) to reduce animal movements at airports. Clearly, it is desirable for airport managers to reduce the attractiveness of airports to wildlife species, particularly those most hazardous to aircraft or airport infrastructure. Reducing the attractiveness of airport resources to hazardous species further enhances the effectiveness of direct control measures by weakening animal tenacity or motivation to use those resources.

Summary

The management of animal movements is directly tied to animal resource needs, including food (Chapter 8), water (Chapter 9), and habitat needs (Chapters 10 and 11). These resource needs are inextricably linked to ecological principles such as natural selection and distribution theories. Animal movements associated with acquiring necessary resources occur at multiple spatial and temporal scales and are linked to physical traits (e.g., body size), biological traits (e.g., reproduction), and ecological traits (e.g., diet, dispersal). Management of wildlife hazards at airports, as well as management of human–wildlife conflicts in general, is frequently most effective through the integration of multiple techniques (Conover 2002). We suggest that current and future practices of wildlife management at airports will benefit from better incorporation of ecological information, including animal movements. This will require an improved understanding the mechanisms responsible for movements of animals hazardous to aircraft, the array of resources (e.g., food and shelter) deemed necessary for persistence of these species, and the spatial constraints or limitations for species acquiring these resources.

LITERATURE CITED

Ayers, C. R., J. L. Belant, and J. J. Millspaugh. 2012. Directness of resource use metrics affects predictions of bear body fat gain. Polar Biology 36:169–176.

Ball, S. A. 2009. Suspending vulture effigies from roosts to reduce bird strikes. Human–Wildlife Conflicts 3:257–259.

Barras, S. C., R. B. Chipman, and R. A. Dolbeer. 2000. Preliminary report on gull behavioral responses to shooting at

John F. Kennedy International Airport, New York. Pages 133–144 in USDA/APHIS/WS National Wildlife Research Center Task 3 Interim Report for the Federal Aviation Administration. DTFA03-99-X-90001. William J. Hughes Technical Center, Atlantic City International Airport, New Jersey, USA.

Basset, A. 1995. Body size-related coexistence: an approach through allometric constraints on home-range use. Ecology 76:1027–1035.

Beason, R. C., J. S. Humphrey, N. E. Myers, and M. L. Avery. 2010. Synchronous monitoring of vulture movements with satellite telemetry and avian radar. Journal of Zoology 282:157–162.

Belant, J. L. 1997. Gulls in urban environments: landscape-level management to reduce conflict. Landscape and Urban Planning 35:245–258.

Belant, J. L., S. K. Ickes, and T. W. Seamans. 1998. Importance of landfills to urban-nesting herring and ring-billed gulls. Landscape and Urban Planning 43:11–19.

Belant, J. L., T. W. Seamans, S. W. Gabrey, and S. K. Ickes. 1993. Importance of landfills to nesting herring gulls. Condor 95:817–830.

Bernhardt, G. E., L. Kutschbach-Brohl, B. E. Washburn, R. B. Chipman, and L. C. Francoeur. 2010. Temporal variation in terrestrial invertebrate consumption by laughing gulls in New York. American Midland Naturalist 163:442–454.

Bernhardt, G. E., Z. J. Patton, L. Kutschbach-Brohl, and R. A. Dolbeer. 2009. Management of bayberry in relation to tree-swallow strikes at John F. Kennedy International Airport, New York. Human–Wildlife Conflicts 3:237–241.

Biondi, K. M., J. L. Belant, J. A. Martin, T. L. DeVault, and G. Wang. 2011. White-tailed deer incidents with U.S. civil aircraft. Wildlife Society Bulletin 35:303–309.

Blackwell, B. F., T. L. DeVault, E. Fernández-Juricic, and R. A. Dolbeer. 2009. Wildlife collisions with aircraft: a missing component of land-use planning for airports. Landscape and Urban Planning 93:1–9.

Blokpoel, H., and B. MacKinnon. 2001. The need for a radar-based, operational bird warning system for civil aviation. Pages 227–231 in Proceedings of the 3rd joint annual meeting. Bird Strike Committee–USA/Canada, Calgary, Alberta, Canada.

Brown, J. S. 1999. Vigilance, patch use and habitat selection: foraging under predation risk. Evolutionary Ecology Research 1:49–71.

Buckley, P. A., and M. G. McCarthy. 1994. Insects, vegetation, and the control of laughing gulls (Larus atricilla) at John F. Kennedy International Airport, New York City. Journal of Applied Ecology 31:291–302.

Caughley, G., and A. R. E. Sinclair. 1994. Wildlife ecology and management. Blackwell Science, Malden, Massachusetts, USA.

Charnov, E. L. 1976. Optimal foraging, the marginal value theorem. Theoretical Population Biology 2:129–136.

Clutton-Brock, T. H. 1989. Female transfer and inbreeding avoidance in social mammals. Nature 337:70–72.

Conover, M. R. 2002. Resolving human–wildlife conflicts: the science of wildlife damage management. CRC Press, Boca Raton, Florida, USA.

Davis, R. A., T. Kelly, R. Sowden, and B. MacKinnon. 2003. Risk-based modeling to develop zoning criteria for land-use near Canadian airports. Proceedings of the 5th joint annual meeting. Bird Strike Committee–USA/Canada, 18–23 August 2003, Toronto, Ontario, Canada.

DeVault, T. L., J. L. Belant, B. F. Blackwell, and T. W. Seamans. 2011. Interspecific variation in wildlife hazards to aircraft: implications for airport wildlife management. Wildlife Society Bulletin 35:394–402.

DeVault, T. L., J. E. Kubel, D. G. Glista, and O. E. Rhodes Jr. 2008. Mammalian hazards at small airports in Indiana: impact of perimeter fencing. Human–Wildlife Conflicts 2:240–247.

Dolbeer, R. A. 2006. Height distribution of birds recoded by collisions with civil aircraft. Journal of Wildlife Management 70:1345–1350.

Dolbeer, R. A. 2011. Increasing trend of damaging bird strikes with aircraft outside the airport boundary: implications for mitigation measures. Human–Wildlife Interactions 5:235–248.

Dolbeer, R. A., J. L. Belant, and J. L. Sillings. 1993. Shooting gulls reduces strikes with aircraft at John F. Kennedy International Airport. Wildlife Society Bulletin 21:442–450.

Dolbeer, R. A., M. Chevalier, P. P. Woronecki, and E. B. Butler. 1989. Laughing gulls at JFK Airport: safety hazard or wildlife resource? Proceedings of the Eastern Wildlife Damage Control Conference 4:37–44.

Dolbeer, R. A., R. B. Chipman, A. L. Gosser, and S. C. Barras. 2003. Does shooting alter flight patterns of gulls? Case study at John F. Kennedy International Airport. Proceedings of the International Bird Strike Committee 2:547–564.

Dolbeer, R. A., and S. E. Wright. 2009. Safety management systems: how useful will the FAA National Wildlife Strike Database be? Human–Wildlife Conflicts 3:167–178.

Dolbeer, R. A., S. E. Wright, and E. C. Cleary. 2000. Ranking the hazard level of wildlife species to aviation. Wildlife Society Bulletin 28:372–378.

Dolbeer, R. A., S. E. Wright, J. Weller, and M. J. Begier. 2010. Wildlife strikes to civil aircraft in the United States 1990–2009. Serial Report 16. U.S. Department of Transportation, Federal Aviation Administration, Office of Airport Safety and Standards, Washington, D.C., USA.

Drent, R., C. Both, M. Green, J. Madsen, and T. Piersma. 2003. Pay-offs and penalties of competing migratory schedules. Oikos 103:274–292.

FAA. Federal Aviation Administration. 2007. Hazardous wildlife attractants on or near airports. Advisory Circular 150/5200-33B. U.S. Department of Transportation, Washington, D.C., USA.

FAA. Federal Aviation Administration. 2010. Airport avian radar systems. Advisory Circular 150/5200-25. U.S. Department of Transportation, Washington, D.C., USA.

Fretwell, S. D. 1972. Populations in seasonal environments. Princeton University Press, Princeton, New Jersey, USA.

Fretwell, S. D., and H. L. Lucas. 1970. On territorial behavior and other factors influencing habitat distribution in birds. I. Theoretical development. Acta Biotheoretica 19:16–36.

Gordon, S., and R. J. White. 2006. Airport canopies become starling roosts—two airport case studies (abstract). Proceedings of the 8th joint annual meeting. Bird Strike Committee–USA/Canada, 21–24 August 2006, St. Louis, Missouri, USA.

Greenwood, P. J. 1980. Mating systems, philopatry and dispersal in birds and mammals. Animal Behaviour 28:1140–1162.

Griffin, C. R., and E. M. Hoopes. 1991. Birds and the potential for bird strikes at John F. Kennedy International Airport. Final Report. National Park Service Cooperative Research Unit, Natural Resources Science Department, University of Rhode Island, Kingston, USA.

Harestad, A. S., and F. L. Bunnell. 1979. Home range and body weight—a reevaluation. Ecology 60:389–402.

Hein, A. M., C. Hou, and J. F. Gillooly. 2012. Energetic and biomechanical constraints on animal migration distance. Ecology Letters 15:104–110.

Heinsohn, R. G. 1991. Slow learning of foraging skills and extended parental care in cooperatively breeding white-winged choughs. American Naturalist 137:864–881.

Hutchinson, G. E. 1957. Concluding remarks. Cold Spring Harbor Symposium on Quantitative Biology 22:415–427.

International Civil Aviation Organization. 1991. Bird control and reduction. Airport Services Manual, Document 9137-AN/898, Part 3. Montreal, Quebec, Canada.

Kacelnik, A. 1984. Central place foraging in starlings (Sturnis vulgaris). 1. Patch residence time. Journal of Animal Ecology 53:283–299.

Kelt, D. A., and D. H. Van Vuren. 1999. Energetic constraints and the relationship between body size and home range area in mammals. Ecology 80:337–340.

Klope, M. W., R. C. Beason, T. J. Nohara, and M. J. Begier. 2009. Role of near-miss bird strikes in assessing hazards. Human–Wildlife Conflicts 3:208–215.

Krebs, C. J. 2001. Ecology: the experimental analysis of distribution and abundance. Benjamin Cummings, San Francisco, California, USA.

Krebs, J. R. 1973. Behavioral aspects of predation. Pages 73–111 in P. P. G. Bateson and P. H. Klopfer, editors. Perspectives in ethology. Plenum, New York, New York, USA.

Kutschbach-Brohl, L., B. E. Washburn, G. E. Bernhardt, R. B. Chipman, and L. C. Francoeur. 2010. Arthropods of a semi-natural grassland in an urban environment: the John F. Kennedy International Airport, New York. Journal of Insect Conservation 14:347–358.

Lindstedt, S. L., D. J. Miller, and S. W. Buskirk. 1986. Home range, time, and body size in mammals. Ecology 67:413–418.

Lonnstedt, O. M., M. I. McCormick, M. G. Meekan, M. C. O. Ferrari, and D. P. Chivers. 2012. Learn and live: predator experience and feeding history determines prey behavioural and survival. Proceedings of the Royal Society B 279:2091–2098.

Mace, G. M., and P. H. Harvey. 1983. Energetics constraints of home range size. American Naturalist 121:120–132.

MacKenzie, D. I., J. D. Nichols, J. A. Royle, K. H. Pollock, L. L. Bailey, and J. E. Hines. 2006. Occupancy estimation and modeling: inferring patterns and dynamics of species occurrence. Academic Press, San Diego, California, USA.

MacKinnon, B. 2006. Avian radar: demonstrated successes and emerging technologies. TP 8240. Airport Wildlife Management Bulletin No. 36. Transport Canada, Ottawa, Ontario, Canada.

Marra, P. P., C. J. Dove, R. Dolbeer, N. Faridah Dahlan, M. Heacker, J. F. Whatton, N. E. Diggs, C. France, and G. A. Henkes. 2009. Migratory Canada geese cause crash of US Airways Flight 1549. Frontiers in Ecology and the Environment 7:297–301.

Martin, J. A., J. L. Belant, T. L. DeVault, B. F. Blackwell, L. W. Burger Jr., S. K. Riffell, and G. Wang. 2011. Wildlife risk to aviation: a multi-scale issue requires a multi-scale solution. Human–Wildlife Interactions 5:198–203.

Millspaugh, J. J., D. C. Kessler, R. W. Kays, R. A. Gitzen, J. H. Schulz, C. T. Rota, C. M. Bodinof, J. L. Belant, and B. J. Keller. 2012. Wildlife radiotelemetry and remote monitoring. Pages 258–283 in N. J. Silvy, editor. The wildlife techniques manual: research. Seventh edition. Johns Hopkins University Press, Baltimore, Maryland, USA.

Montoney, A. J., and H. C. Boggs. 1993. Effects of a bird hazard reduction force on reducing bird/aircraft strike hazards at the Atlantic City International Airport, NJ. Eastern Wildlife Damage Control Conference 6:59–66.

Nohara, T. J., R. C. Beason, and P. Weber. 2011. Using radar cross-section to enhance situational awareness tools for airport avian radars. Human–Wildlife Interactions 5:210–217.

Olsson, O., J. S. Brown, and K. L. Helf. 2008. A guide to central place effects in foraging. Theoretical Population Biology 74:22–33.

Orians, G. H., and N. W. Pearson. 1979. On the theory of central place foraging. Pages 154–177 in D. J. Horn, R. D. Mitchell, and G. R. Stairs, editors. Analyses of ecological systems. Ohio State University Press, Columbus, USA.

Schafer, L. M., J. L. Cummings, J. A. Yunger, and K. E. Gustad. 2002. Efficacy of translocation of red-tailed hawks from airports (abstract). Proceedings of the 4th joint annual meeting. Bird Strike Committee–USA/Canada, 21–24 October 2002, Sacramento, California, USA.

Schoener, T. W. 1971. Theory of feeding strategies. Annual Review of Ecology and Systematics 2:369–404.

Schumacher, A. K., B. E. Washburn, and J. P. Hart. 2008. Would you please move? Translocation of immature bald eagles to reduce bird strike risk (abstract). Proceedings of the 10th joint annual meeting. Bird Strike Committee–USA/Canada, 18–21 August 2008, Orlando, Florida, USA.

Servoss, W., R. M. Engeman, S. Fairaizl, J. L. Cummings, and N. P. Groninger. 2000. Wildlife hazard assessment for Phoenix Sky Harbor International Airport. International Biodeterioration and Biodegradation 45:111–127.

Silva, M., and J. A. Downing. 1995. The allometric scaling of density and body mass: a nonlinear relationship for terrestrial mammals. American Naturalist 145:704–727.

Sirot, E. 2010. Should risk allocation strategies facilitate or hinder habituation to nonlethal disturbance in wildlife? Animal Behavior 80:737–743.

Sirot, E., and F. Touzalin. 2009. Coordination and synchronization of vigilance in groups of prey: the role of collective detection and predators' preference for stragglers. American Naturalist 173:47–59.

Thornton, A., and T. Clutton-Brock. 2011. Social learning and the development of individuals and group behavior in mammal societies. Philosophical Transactions of the Royal Society B 366:978–987.

Tomkiewicz, S. M., M. R. Fuller, J. G. Kie, and K. K. Bates. 2010. Global positioning system and associated technologies in animal behavior and ecological research. Philosophical Transactions of the Royal Society B 365:2163–2176.

van Wijk, R. W., A. Kölzsch, H. Kruckenberg, B. S. Ebbinge, G. J. D. M. Müskens, and B. A. Nolet. 2012. Individually tracked geese follow peaks of temperature acceleration during spring migration. Oikos 121:655–664.

Wakefield, E. D., R. A. Phillips, J. Matthiopoulos, A. Fukuda, H. Higuchi, G. J. Marshall, and P. N. Trathan. 2009. Wind field and sex constrain the flight speeds of central-place foraging albatrosses. Ecological Monographs 79:663–667.

Waser, P. M. 1996. Patterns and consequences of dispersal in gregarious carnivores. Pages 267–295 in J. L. Gittleman, editor. Carnivore behavior, ecology, and evolution. Cornell University Press, Ithaca, New York, USA.

Waser, P. M., and W. T. Jones. 1983. Natal philopatry among solitary mammals. Quarterly Review of Biology 58:355–390.

Washburn, B. E., G. E. Bernhardt, and L. A. Kutschbach-Brohl. 2011. Using dietary analysis to reduce the risk of wildlife–aircraft collisions. Human–Wildlife Interactions 5:204–209.

Washburn, B. E., G. E. Bernhardt, L. Kutschbach-Brohl, R. B. Chipman, and L. C. Francoeur. 2013. Foraging ecology of four gull species in a coastal-urban interface. Condor 115:67–76.

Washburn, B. E., B. N. Haslun, M. S. Lowney, and S. E. Tennis. 2009. Shooting gulls to reduce strikes with aircraft at John F. Kennedy International Airport, 1991–2008. Special Report for the Port Authority of New York and New Jersey. U.S. Department of Agriculture, Wildlife Services, National Wildlife Research Center, Sandusky, Ohio, USA.

Washburn, B. E., M. S. Lowney, and A. L. Gosser. 2012. Historical and current status of Laughing Gull breeding in New York State. Wilson Journal of Ornithology 124:525–530.

Washburn, B. E., and T. J. Olexa. 2011. Assessing BASH risk potential of migrating and breeding osprey in the mid-Atlantic Chesapeake Bay region. Final Project Report. U.S. Department of Defense, Legacy Resources Management Program, Arlington, Virginia, USA.

Wilson, R. P., F. Quintana, and V. J. Hobson. 2012. Construction of energy landscapes can clarify the movement and distribution of foraging animals. Proceedings of the Royal Society B 279:975–980.

York, D. L., J. L. Cummings, R. M. Engeman, and K. L. Wedemeyer. 2000a. Hazing and movements of Canada geese near Elmendorf Air Force Base in Anchorage, Alaska. International Biodeterioration and Biodegradation 45:103–110.

York, D. L, J. L. Cummings, and K. L. Wedemeyer. 2000b. Movements and distribution of radio-collared Canada geese in Anchorage, Alaska. Northwestern Naturalist 81:11–17.

13
SIDNEY A. GAUTHREAUX JR.
PAIGE M. SCHMIDT

Radar Technology to Monitor Hazardous Birds at Airports

Bird strikes are the most common wildlife hazard to aviation safety (Dolbeer et al. 2000). Advances in habitat management at airports through the elimination and reduction of attractants, in combination with hazing and lethal control, have reduced avian hazards <152 m (500 feet) above ground level. Bird strikes above this altitude, however, are beyond the limits of traditional wildlife control techniques (Dolbeer 2011). Traditional avian survey methods used to monitor birds at airports (Cleary and Dolbeer 2005; Chapter 14) often fail to provide essential information on local bird activity and migration at higher altitudes, hazardous bird use of attractants near airports, and bird activity at night—information that could be provided by the strategic use of radar technology at and near airports (Dolbeer 2006).

Radar in its simplest form is transmission of a pulse of energy, reflection of a portion of the transmitted energy by a target, and reception of the returned energy by a receiver (Eastwood 1967). The time delay between transmission and reception is used to determine the range of a given target. Radar can generally provide each target's bearing, flight speed, and altitude (depending on the type of radar antenna), having been originally developed to track enemy aircraft during World War II (Lack and Varley 1945). Early users of radar discovered that it could detect and track birds, commonly referred to as "angels" (Lack and Varley 1945). Biologists have exploited the ability of radar to detect and track birds for several decades, including radars located at airports. The first major, coordinated use of a group of radars to study bird movements over a large region was initiated in Canada in 1964 to address bird collisions with aircraft (Eastwood 1967). This was soon followed by additional evaluation (Gauthreaux 1972) and the recognition of radar's potential as an effective tool for providing early warnings of birds hazardous to aircraft (Blokpoel 1976). Some uses of radar relied on co-opting existing radar technology for bird detection (Blokpoel 1976; see the following sections), but radar technology has been recently adapted to detect birds in the airport environment (Federal Aviation Administration 2010). Different types of radar operate at different spatial scales (i.e., resolution and extent) and can be used to gather different types of data on bird movements in the atmosphere. In this chapter, for each type of radar used currently in ornithology, we provide information on technical capabilities and limitations, types of data that can be acquired, and how they can or are being used to detect hazardous birds. We also suggest how this technology could complement existing management practices (e.g., habitat modification) to reduce the risk of bird collisions with aircraft.

Radar Sensors Used in Ornithology

Tracking Radar

Tracking radar has been used to gather detailed information on the flight paths and speeds of individual migrating and foraging birds (Bruderer and Steidinger 1972, Griffin 1972, Able 1977, Kerlinger 1982, Larkin and Frase 1988, Bruderer 1999, Bäckman and Alerstam 2003). Small, military tracking radars (40–200 kW) have narrow beams (e.g., <1°–3°) and can detect individual targets from 0.1 to 6.0 km (<0.1 to 3.8 miles). Radar "locks on" a target, and the radar antenna follows it until the target moves too far away and the return signal is lost, or until another target enters the beam at the same range, causing the radar to switch to the new target. The position of the target in three-dimensional space and the strength of the reflected signal are digitally recorded continuously for subsequent analysis. Tracking radar can provide information that could prove useful to study behavioral responses of birds to approaching aircraft. Tracking radar also can provide information on wingbeat patterns (Bruderer et al. 2010), data that could be used to identify or classify targets to species or groups. Although the number of targets sampled may be limited, the beam can be rotated in a horizontal surveillance mode to sample migrating birds over a greater area (Bruderer et al. 1995). It is also possible to operate tracking radar in a fixed-beam mode and to monitor birds passing through the stationary beam (Larkin and Eisenberg 1978, Schmaljohann et al. 2008).

Weather Surveillance Radar

In the USA, weather surveillance radar (WSR) has been used to study bird movements and bat roosts since the late 1950s. In the early 1990s the WSR-88D (also known as NEXRAD, or NEXt Generation RADar) replaced the older WSR-57, WSR-74S, and WSR-74C radars in the national network. There are now 159 sites throughout the USA and overseas locations (Fig. 13.1). WSR-88D technology is more advanced than technology in older WSRs (Crum and Alberty 1993, Crum et al. 1993, Klazura and Imy 1993), and the improved sensitivity enhanced detection of weak targets such as birds, bats, and insects (Larkin 1984). These powerful (500 kW) and sensitive (45.8 dB) S-band (10-cm-wavelength) Doppler WSRs have a 1.0° beam, and when the beam is tilted 0.5° above the horizontal, the radar can detect concentrations of biological targets up to 240 km (149 miles) away and intense precipitation at a maximum range of 460 km (286 miles). The antenna of the WSR-88D is computer controlled and repeatedly scans the atmosphere through a sequence of predefined elevation angles, antenna rotation rates, and pulse charac-

Fig. 13.1. Locations of the 159 WSR-88D stations throughout the USA and territories. Map available at http://radar.weather.gov/index.htm

teristics (i.e., volume coverage patterns), depending on the radar's mode of operation. Two operational modes exist—a precipitation mode and a clear-air mode—and selection of an operational mode is closely related to the detected coverage of precipitation. The WSR-88D is sensitive enough to detect birds, bats, and concentrations of insects in precipitation mode. When no precipitation is detected, the radar operates in clear-air mode and samples the same volume of airspace more slowly, making it possible to detect the reflected energy from small objects such as insects and even dust and smoke particles. Since August 2008, the resolution of the reflectivity data has increased to 0.25 km (820 feet) by 0.5° to match the velocity data, and velocity data were extended from 230 to 300 km (143 to 186 miles). By May 2013, all WSR-88D stations will have been upgraded to dual-polarization technology that will add three new base products (differential reflectivity, correlation coefficient, and specific differential phase) that will aid meteorologists and biologists in identifying and quantifying radar returns from weather and biological targets in the atmosphere (Zrnic and Ryzhkov 1998, Gauthreaux et al. 2008).

Biological Data Provided by the WSR-88D

The WSR-88D can readily detect aerial biological targets, and several investigators have used it to study bird migration (Gauthreaux and Belser 1998, 1999b, 2003a; Diehl and Larkin 2005), bird roosts (Russell and Gauthreaux 1998, Russell et al. 1998), bat colonies (McCracken 1996, McCracken and Westbrook 2002, Horn and Kunz

2008), and concentrations of insects aloft (Westbrook and Wolf 1998). The WSR-88D can be used to quantify the number of birds in migration aloft (Gauthreaux and Belser 1998, 1999*a*; Black and Donaldson 1999; Diehl et al. 2003, Gauthreaux et al. 2008) and has been used to study regional bird migration patterns on the northern Gulf Coast (Gauthreaux and Belser 1999*b*), in the Great Lakes region (Diehl et al. 2003), across the USA–Mexico borderlands region (Felix et al. 2008), and at a continental scale (Gauthreaux et al. 2003). Digital data files can be obtained from the WSR-88D archives at the National Climatic Data Center in Asheville, North Carolina, USA, and detailed methods of analyzing data from the WSR-88D can be found in Gauthreaux and Belser (2003*a*), Diehl and Larkin (2005), Gauthreaux et al. (2008), and Buler and Diehl (2009).

Within 120 km (75 miles) of the radar, WSR-88D can be used to delimit important migration stopover areas by measuring bird density (birds per cubic kilometer) in the beam as they begin a migratory movement (Gauthreaux and Belser 2003*b*, Bonter et al. 2009, Buler and Diehl 2009). Within minutes of the onset of nocturnal migration, the distribution and density of echoes in the radar beam can provide information on geographical ground sources of the migrants, and satellite imagery can be used to identify the topography and habitat type that characterize these areas (Gauthreaux and Belser 2003*b*). Bird stopover areas have been mapped using the displays of the WSR-88D for areas in eastern Louisiana and southern Mississippi (Buler and Diehl 2009), for radar sites around the Great Lakes (Bonter et al. 2009), and for several sites at and near military installations (Fischer et al. 2012). At ranges >120 km, this approach can be used to delimit locations of postbreeding and nocturnal roost sites of birds such as purple martins (*Progne subis*; Fig. 13.2), as well as to quantify the density of birds (Russell and Gauthreaux 1998) and bats (Horn and Kunz 2008).

The greatest limitation of the WSR-88D for use in biological studies has been the size of the radar's legacy pulse volumes (1° × 1 km), which increases with increasing distance. This corresponding growth prohibits gathering information on small, individual targets and combines the return from several different types of targets into one pulse volume. The upgrade to superresolution should improve this shortcoming, but resolution cells (0.5° × 250 m) will still be sampling

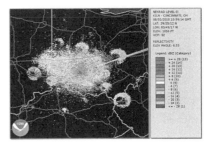

Fig. 13.2. Display of the WSR-88D radar at Cincinnati, Ohio, USA, at 1039 GMT on 2 August 2010. The circles show Purple Martins (*Progne subis*) departing from overnight roosts. The strobe is from the rising sun, which emits microwaves similar to those emitted by the radar. The density of birds can be estimated from the reflectivity scale (in decibels relative to *Z*, or dBZ) on the right.

Fig. 13.3. Locations and station codes of the 45 terminal Doppler weather radar units in the USA. The units are located near airports to monitor wind shear and severe weather.

a large volume of atmosphere. Because the lowest antenna scan is at an angle of 0.5° above the horizontal, it is commonplace for low-flying targets to go undetected because they are below radar coverage.

Terminal Doppler Weather Radar

Although terminal Doppler weather radar (TDWR) has not been assessed adequately for its ability to detect migrating birds, its operational characteristics suggest it should be an excellent sensor for that purpose (Istok et al. 2008). TDWR was developed for the Federal Aviation Administration in the early 1990s to detect real-time wind shear and high-resolution precipitation data, and as of 2009, 45 units were deployed near major airports across the USA (Fig. 13.3). The radar operates at the *C* band or 5-cm wavelength (5,600–5,650 MHz) and has a peak power of 250 kW. Antenna beam width is 0.55°, and the antenna completes twenty-three 360° sweeps every 6 min in severe/hazardous mode. Reflectivity of targets can be measured to 460 km distant while Doppler (radial) velocity of targets can be measured to 89 km (55 miles). Although similar in operation to the WSR-88D, the resolution of TDWR is greater, and TDWR antennae can scan below an angle of 0.5° above the horizontal, providing information on bird activity at the scale of an airfield.

High-Resolution Marine Surveillance Radars

Casement (1966) was one of the first to use marine surveillance radar on a ship to study bird migration, and interest in using marine radar to study bird movements subsequently increased (Williams et al. 1972, Williams 1984). Because of the relatively low cost of marine surveillance radar, this technology has been used extensively for bird detection at airports (e.g., MacKinnon 2006) and for environmental impact studies (e.g., National Academy of Sciences 2007).

Technical Specifications

The following radar characteristics are known to influence the results obtained from radar studies of bird movements:

- Transmitter power (e.g., 5, 10, 25, 50, or 60 kW)
- Frequency or wavelength
- Pulse length and corresponding pulse repetition frequencies
- Antenna beam characteristics
- Antenna rotation speed
- Tuning of the receiver
- Magnetron or solid state
- Gain setting
- Range setting
- Ground and sea- and rain-clutter settings
- Beam-brilliance setting

Most of the small, mobile radars used to monitor bird movements have been low-powered (5–60 kW) marine-surveillance radars of 3- or 10-cm wavelengths and are commonly referred to as "avian radars." The transmitter power of the avian radar should be as high as possible (≥ 25 kW) to maximize resolution and sensitivity. Long pulse lengths enhance detectability but

Fig. 13.4. Antenna configurations commonly used in avian radar systems are designed to detect and track hazardous birds at airports: (*left*) slotted arrays for horizontal and vertical scanning, and (*right*) parabolic dish antennas.

have lower resolution, whereas short pulse lengths increase resolution with decreased detectability. The greater the transmitter power, the greater the cost, but a 50-kW radar operating on short pulse will produce superior results for bird detection than a 10-kW unit operating on short pulse. Marine radars can be purchased in either of two wavelengths—3 cm (X band) or 10 cm (S band)—and there is considerable debate among users of these two radar types regarding which one is best. Both have been used to study bird movements aloft, but no published study has compared them at the same location and under similar weather conditions. Precipitation attenuates 3-cm wavelengths considerably more than it does 10-cm signals (LGL Environmental Research Associates 2000); consequently, intense precipitation will greatly decrease the chances of detecting targets using 3-cm radar. Regardless of wavelength, small-target detection during heavy precipitation is not likely.

In typical horizontal surveillance mode (Fig. 13.4), the radar beam samples 20–25° of vertical airspace and has a horizontal (azimuth) resolution of 1.0–2.3°. These radars can detect movements of individual birds out to several kilometers, and the exact range of detection depends on the power of the radar and the size of the birds. In horizontal surveillance mode the altitude of a target cannot be measured because of the vertical extent of the radar beam. To address this limitation, the radar transmitter/receiver and array antenna can be tilted 90° so that the sweep of the antenna is vertical (Fig. 13.4). In vertical surveillance mode (20° in horizontal and 1° in vertical) the altitude of a target can be accurately measured, but the 20° sweep from horizon through zenith to opposite horizon is restricted to one axis. Because of the axial surveillance pattern of vertical scanning radar, targets moving parallel to the axis of the sweep show true ground speeds as they are tracked. Targets moving at increasing angles to the axis of antenna sweep show reduced ground speeds, and targets moving perpendicular to the sweep have zero ground speeds and appear stationary (if they have enough detections to be tracked).

Some investigators have used a single radar for horizontal and vertical surveillance (Harmata et al. 2003), whereas others have used two radars, one each for horizontal and vertical surveillance (Harmata et al. 1999). An alternative design replaces the open-array antenna with a rotating, parabolic antenna (Fig. 13.4) that projects a narrow, conical (e.g., 2.5–4.0°) beam that can be raised or lowered (Gauthreaux and Belser 2003b, Nohara et al. 2005). When the conical beam is elevated in the horizontal surveillance mode, the altitude of an echo is a trigonometric function of the range of the echo and the angle of antenna tilt. When the antenna is elevated 30°, for example, the altitude of a target is one-half of the range. The advantage of the parabolic dish is that information on range and altitude can be obtained for each echo. The open-array antenna samples a greater volume of airspace, but the altitude of a target in the vertical scan cannot be associated with the track of a target in the horizontal scan. The parabolic antenna samples a smaller volume of atmosphere but has higher gain, and three-dimensional information on each target can be measured. Antenna rotation speed is dependent on gear configuration and is usually ~24 revolutions per minute. Higher rotation rates are possible and provide additional detections for tracking a target, but a target receives fewer radar pulses per detection at higher rotation speeds.

When tuned properly, avian radars can detect individual birds within 2–3 km (1–2 miles) and large flocks of large birds to 10–14 km (6–9 miles; Gauthreaux and Belser 2003b). Desholm et al. (2004) reported that European thrushes (*Turdus* spp.) can be detected with 10- and 12-kW units to 6.0 km from the radar, and with a 25-kW, 3-cm-wavelength radar in clear weather, an 800-g duck can be detected to 2.2 km (1.4 miles) for short pulse and 3.2 km (2 miles) for long pulse, whereas

the maximum range of detection for small passerines is 800–1,000 m (2,625–3,281 feet). They also report that a 500-g pigeon-like target can be detected at 4.0 km (2.5 miles) for short pulse and 5.5 km (3.4 miles) for long pulse with a 60-kW, 10-cm-wavelength radar in clear weather. Other radar ornithologists have found that a 12-kW radar (with an open-array antenna) can routinely detect flocks of waterfowl to 5.6 km (3.5 miles), individual hawks to 2.3 km (1.4 miles), and single, small passerines to 1.2 km (0.75 miles; Cooper et al. 1991, 2004). Range discrimination depends on pulse length used, and with short pulse lengths, minimum detectable range can be as close as 20–30 m (66–98 feet); however, not all marine radars detect biological targets equally (J. Kube, Institut für Angewandte Ökologie, personal communication, 2005).

Radar Performance and Data Quality

Technical limitations can affect the quality of data gathered by avian radars. The aspect of the bird relative to the radar beam affects the amount of energy reflected back to the radar receiver, such that head-on and tail-on detections have smaller radar cross sections than broadside detections. The radar cross section (RCS) of a bird is dependent on properties such as size, mass, and water content, and is independent of the range of the target relative to the radar. To determine RCS, intensity of a target's radar signal must first be measured and then corrected for wave propagation effects (Nohara et al. 2011). Because RCS is size dependent, it can be used to estimate the sizes of birds in radar tracks. The position of the bird in the radar beam is another important consideration. Radar beam width is defined as the angle where the energy at the center of the beam is reduced by one-half (or –3 dB). If two identical targets are located at the same range, the target at the edge of the radar beam will produce a weaker echo than the target at the center of the beam. Similarly, a strong target outside the radar beam can be detected as a weak target. The latter problem is amplified when using an array antenna (20–25°), because the power loss beyond the half-power point is more gradual than power loss in high-gain pencil beams. When birds fly low to the ground, they often go undetected by marine radar. In a review of bird migration studies with radar, Bruderer (1997) reported that marine radar missed about 40% of

Fig. 13.5. An image generated from digitally processed data from a Furuno 2155-BB radar with a parabolic dish (4° beam width) elevated 30° above the horizontal showing tracks of nighttime migrating birds in fall over coastal Maryland, USA. Tracks are series of target detections, and the current position and heading of the target are indicated with "lollipop" symbols. Source: Tim J. Nohara, Accipiter Radar Technologies Inc.

from the receiver and then process that data using proprietary algorithms. The algorithms mask ground clutter and use the data from target detections to generate target tracks that are reported either in spreadsheet format with information for every detection in a track (e.g., reflectivity, range, size of echo, and speed) or as plots showing target tracks. Automatic digital processing is extremely fast and eliminates the potential bias associated with manual data extraction and processing, but automatic processing algorithms also have shortcomings and must be evaluated carefully to expose systematic biases in the algorithms. Algorithms that require a certain number of detections before tracking begins could potentially exclude fast targets that produce fewer than the required number of detections. Hundreds of targets can be tracked at once, but as the number of targets increases, so does the possibility that tracking algorithms may switch between nearby targets and treat two different tracks as one. When the radar is recording a large number of detections from rain or waves, the tracking algorithms will produce false tracks that satisfy the algorithms, but they are not real bird tracks. There is clearly a need to carefully ground truth the reports of data from digitally processed radar return, but few published studies have done so.

Radar Validation

The determination of the number of targets per echo and the identification of the source of the echo (e.g., birds, or bats, or insects) on avian radars can be problematic. One cannot generally discriminate an individual target from a tight cluster of targets, because a single large target may produce the same echo as a tight group of smaller targets. It is nearly impossible to discriminate echoes from similarly sized birds and bats on the basis of echo characteristics, and flight behavior may be similar between foraging bats and nocturnally foraging birds (e.g., nighthawks) and between migrating bats and migrating birds engaged in linear flight. This uncertainty has led investigators to refer to the sources of echoes in radar studies as "biological targets." It is possible to characterize targets by their airspeed if one knows the speed and direction of the wind at the altitude where a target is detected. Once the airspeed of the target is calculated, it can be assigned to categories of bird types based on airspeed

birds flying below 50 m (164 feet), but when birds were flying above 50 m, only 8% were undetected.

Return from ground objects produces clutter in radar displays, and if the ground-clutter return signals are strong and extensive, return from birds will be obscured. Although algorithms have been developed to filter clutter, in many instances, bird detection over areas where clutter has been removed is reduced, particularly when the targets are small, single birds. Constant false-alarm rate processing can be used to detect return signals from moving targets in clutter, but the clutter threshold must be consistent between scans, a requirement that is likely to be violated.

Two methods of collecting and processing avian radar data exist. At first, investigators manually extracted the echo data from the radar display (or a digital image of the display; Fig. 13.5) and then performed analyses to compute descriptive statistics. Manual data extraction is labor-intensive, time-consuming, and presents the possibility of bias. More recently, radars with digital processors have been used to gather raw radar data

(Harmata et al. 1999). In some instances the flight behavior of a target may offer clues to its identity (e.g., circling of a raptor in a thermal), but claims of target identification based on size of target (number of pixels) are likely incorrect. Many attempts to statistically link echo characteristics to the identity of hundreds of known targets have shown no significant relationship (O. Hüppop, Institut für Vogelforschung Vogelwarte Helgoland, personal communication, 2006). The best means of identifying the sources of radar echoes involve simultaneous visual observations during the day with binoculars, telescope, or high-definition video, as well as the use of thermal imaging (Gauthreaux and Livingston 2006) and infrared devices (Plissner et al. 2006) at night. Because light may attract birds, insects, and bats that feed on insects, techniques that require illuminating targets should be avoided. Radar targets can be verified only when they are within the range limits of the method used for verification.

Use of Avian Radar Data

MacKinnon (2006) compiled information on small radars used to detect, monitor, and quantify bird movements that pose a threat to aircraft. Avian radars have been deployed at both military (e.g., Klope et al. 2009, Beason et al. 2010a, Coates et al. 2011) and civil airfields (Federal Aviation Administration 2010), although inherent differences between the two types of airfields will determine how avian radar data can be applied to reduce the risk of bird strikes. Civil aircraft strike most birds near airports in the approach and departure corridor (Dolbeer et al. 2009), whereas military aircraft have the additional risk of striking birds during low-altitude, high-speed training flights (Zakrajsek and Bissonette 2005). Civil airfields rely on mitigation of wildlife hazards to reduce bird-strike risks (i.e., habitat management, harassment, and lethal control; Cleary and Dolbeer 2005), whereas military airfields also use bird avoidance models to schedule low-level training flights during periods with low strike risk (Zakrajsek and Bissonette 2005). Avian radars could provide substantial data (e.g., local bird use and migration at higher altitudes, use of attractants near airports, and nocturnal activity; Dolbeer 2011) for use in Wildlife Hazard Management Plans (Cleary and Dolbeer 2005), trend analysis, and real-time warnings both to air operations

staff and air traffic control personnel (Blokpoel and MacKinnon 2001, Kelly et al. 2007). Avian radar data can also be used to develop local bird-strike risk management models specific to civil or military airfields (Coates et al. 2011) and as a metric to assess bird–aircraft collision mitigation strategies (Klope et al. 2009).

Many professionals involved in reducing bird–aircraft collisions believe that high-resolution marine radar or newly developed avian radar will be an important component of future bird-strike mitigation systems. But questions remain regarding detection and tracking capabilities, reliability, and proper use of avian radar systems at airports (Weber et al. 2005). The use of avian radar is relatively new at civilian airports; Federal Aviation Administration (2010) provides guidelines for selecting and deploying avian radar systems. These guidelines are relatively flexible because of the variability of available hardware and software, as well as the hazards and geography specific to each airfield that influence system performance.

Recent Developments in Avian Radar

Existing shortcomings of horizontal surveillance and vertical avian radar systems have stimulated development of new radar configurations and entirely new systems. Some developers have moved from a two-radar system to a single-radar system with single or dual antennas. Others have changed the type of radar used for vertical scanning, or are in the process of developing Doppler marine radars. The sweep axis of vertically scanning radar can be shifted by 20° every 3 min, resulting in 72 vertical scans for each 20° sector, and nine sectors are sampled in 27 min. This mode of operation generates 360° coverage within 27 min and eliminates the sampling bias of collecting data while operating on only one axis. In addition, a stationary thermal imaging camera (TIC) can be mounted next to the transmitter/receiver unit and pointed vertically to sample targets passing through the fixed 20° field of view of the TIC. The TIC data can be used to identify the sources of the radar echoes. This configuration also can be shifted 90°. A dual-beam antenna radar can be built with two standard dish antennas (4° beam width; Beason et al. 2010b). The radar connection can be alternated between the two dishes from one pulse to the next, and the data stream tagged to indicate which antenna was active for each pulse. The beam patterns are identical for both dishes (one dish set to 7° elevation and the second at 11° with overlap at beam half-power points). When the dual-beam antenna radar becomes operational, three-dimensional systems will have altitude computations embedded in the real-time processor. A two-radar system can combine horizontal scanning marine radar (*X* band or *S* band) and frequency-modulated continuous wave radar (two antennas) used to track a bird and to measure altitude and wingbeat pattern and frequency (Borst 2009). An avian radar is also available that uses a monostatic pulse radar and Doppler-like processing to determine target velocities. It processes received echoes in a bank of narrowband, coherently integrating filters that resolve targets within particular velocity bands. Some new avian radar systems are no longer based on marine surveillance radars, such as the solid-state, mobile surveillance and target acquisition radar, which uses an electronic beam (*L* band) to scan 360° with no moving parts and provides three-dimensional target information. Finally, avian radar systems can also be connected to an apparatus that automatically hazes birds when detected by the radar.

Summary

The use of radar to study bird behavior has a long history that began during the early years of military radar. The modernization of the national weather radar system expanded radar ornithology to include studies of bird movements at the regional and continental scales. Adaptation of small, mobile marine radars led most recently to the availability of bird movement data specific to individual airfields. Radar data on migratory bird movements are currently being used to provide air operations personnel (e.g., flight schedulers, planners, and pilots) with near-real-time warnings of hazardous flight conditions caused by large movements of migratory birds. Avian radar data can also be used to develop local bird-strike risk management models specific to a civil or military airfield and to assess bird–aircraft collision mitigation strategies. Additionally, these data can be used in trend analysis and have been cautiously proposed to provide real-time warnings, both to airport operations staff and air traffic control personnel, although the feasibility of the latter application is highly debated (Nohara 2009).

Several reports have indicated that a radar beam pointed directly at a flock of birds resulted in dispersal behavior (Eastwood 1967). These accounts focused mainly on the effect on flight; however, other behavioral effects (e.g., predator detection, foraging ability, and ability to locate cached food) could be influenced by incident microwave radiation and thus could potentially influence survival and fitness. Research is now underway to determine whether microwave radiation emitted by various forms of radar technology influences bird behavior or has potential as a deterrent device (E. Fernández-Juricic, Purdue University, unpublished data).

The broad spectrum of available and developing technology will influence the quality, quantity, and application of radar data to reduce bird–aircraft collisions. The limitations of the data must be acknowledged and additional studies conducted to evaluate appropriate uses of information provided by this technology. The novelty of information collected by radars will not compensate for bias inherent in poor methodology, or for failure to understand how the hardware and software specific to each application influence the information provided.

LITERATURE CITED

Able, K. P. 1977. The flight behavior of individual passerine nocturnal migrants: a tracking radar study. Animal Behaviour 25:924–935.

Bäckman, J., and T. Alerstam. 2003. Orientation scatter of free-flying nocturnal passerine migrants: components and causes. Animal Behaviour 65:987–996.

Beason, R. C., J. S. Humphrey, N. E. Myers, and M. L. Avery. 2010a. Synchronous monitoring of vulture movements with satellite telemetry and avian radar. Journal of Zoology 282:157–162.

Beason, R. C., P. Weber, and T. J. Nohara. 2010b. Color vision as a model for precise altitude determination for avian radar. Proceedings of the 29th annual meeting. International Bird Strike Committee, 21–24 September 2010, Cairns, Queensland, Australia.

Black, J. E., and N. Donaldson. 1999. Comments on "Display of bird movements on the WSR-88D: Patterns and quantification." Weather and Forecasting 14:1039–1040.

Blokpoel, H. 1976. Bird hazards to aircraft: problems and prevention of bird/aircraft collisions. Clarke, Irwin, Ottawa, Ontario, Canada.

Blokpoel, H., and B. MacKinnon. 2001. The need for a radar-based, operational bird warning system for civil aviation (abstract). Proceedings of the 3rd joint annual meeting.

Bird Strike Committee–USA/Canada, 27–30 August 2001, Calgary, Alberta, Canada.

Bonter, D. N., S. A. Gauthreaux, and T. M. Donovan. 2009. Characteristics of important stopover locations for migrating birds: remote sensing with radar in the Great Lakes Basin. Conservation Biology 23:440–448.

Borst, A. 2009. Revolutionizing airport bird strike prevention with the ROBIN Lite Search & Track Bird Radar (abstract). Proceedings of the Bird Strike North America conference. Bird Strike Committee–USA/Canada, 14–17 September 2009, Victoria, British Columbia, Canada.

Bruderer, B. 1997. The study of bird migration by radar. Part 2: major achievements. Naturwissenschaften 84:45–54.

Bruderer, B. 1999. Three decades of tracking radar studies on bird migration in Europe and the Middle East. Pages 107–141 in Y. Leshem, Y. Mandelik, and J. Shamoun-Baranes, editors. Migrating birds know no boundaries. International Center for the Study of Bird Migration, Latrun, Israel.

Bruderer, B., D. Peter, A. Boldt, and F. Liechti. 2010. Wing-beat characteristics of birds recorded with tracking radar and cine camera. Ibis 152:272–291.

Bruderer, B., and P. Steidinger. 1972. Methods of quantitative and qualitative analysis of bird migration with tracking radar. Pages 151–167 in S. R. Galler, K. Schmidt-Koenig, G. J. Jacobs, and R. E. Belleville, editors. Animal orientation and navigation. NASA SP-262. National Aeronautics and Space Administration, Washington, D.C., USA.

Bruderer, B., T. Steuri, and M. Baumgartner. 1995. Short-range high-precision surveillance of nocturnal migration and tracking of single targets. Israel Journal of Zoology 41:207–220.

Buler, J. J., and R. H. Diehl. 2009. Quantifying bird density during migratory stopover using weather surveillance radar. IEEE Transactions on Geoscience and Remote Sensing 47:2741–2751.

Casement, M. B. 1966. Migration across the Mediterranean observed by radar. Ibis 108:461–491.

Cleary, E. C., and R. A. Dolbeer. 2005. Wildlife hazard management at airports, a manual for airport personnel. Second edition. U.S. Department of Transportation, Federal Aviation Administration, Office of Airport Safety and Standards, Washington, D.C., USA.

Coates, P. S., M. L. Casazza, B. J. Halstead, J. P. Fleskes, and J. A. Laughlin. 2011. Using avian radar to examine relationships among avian activity, bird strikes, and meteorological factors. Human–Wildlife Interactions 5:249–268.

Cooper, B. A., R. H. Day, R. J. Ritchie, and C. L. Cranor. 1991. An improved marine radar system for studies of bird migration. Journal of Field Ornithology 62:367–377.

Cooper, B. A., T. J. Mabee, A. A. Stickney, and J. E. Shook. 2004. A visual and radar study of spring bird migration at the proposed Chautauqua Wind Energy Facility, New York. Prepared for Chautauqua Windpower. ABR, Forest Grove, Oregon, USA.

Crum, T. D., and R. L. Alberty. 1993. The WSR-88D and the WSR-88D operational support facility. Bulletin of American Meteorological Society 74:1669–1687.

Crum, T. D., R. L. Alberty, and D. W. Burgess. 1993. Recording, archiving, and using WSR-88D data. Bulletin of the American Meteorological Society 74:645–653.

Desholm, M., A. D. Fox, and P. D. Beasley. 2004. Best practice guidance for the use of remote techniques for observing bird behavior in relation to offshore wind farms. Report Prepared for Collaborative Offshore Wind Research into the Environment Consortium. National Environmental Research Institute, Rønde, Denmark, and QinetiQ, Malvern Technology Centre, Worcestershire, United Kingdom.

Diehl, R. H., and R. P. Larkin. 2005. Introduction to the WSR-88D (NEXRAD) for ornithological research. Pages 876–888 in C. J. Ralph and T. D. Rich, editors. Bird conservation implementation and integration in the Americas: proceedings of the 3rd international partners in flight conference. Volume 2. General Technical Report PSW-GTR-191. U.S. Department of Agriculture, Forest Service, Pacific Southwest Research Station, Albany, California, USA.

Diehl, R. H., R. P. Larkin, and J. E. Black. 2003. Radar observation of bird migration over the Great Lakes. Auk 120:278–290.

Dolbeer, R. A. 2006. Height distribution of birds as recorded by collisions with civil aircraft. Journal of Wildlife Management 70:1345–1350.

Dolbeer, R. A. 2011. Increasing trend of damaging bird strikes with aircraft outside the airport boundary: implications for mitigation measures. Human–Wildlife Interactions 5:235–248.

Dolbeer, R. A., S. W. Wright, and E. C. Cleary. 2000. Ranking the wildlife hazard level of wildlife species to aviation. Wildlife Society Bulletin 28:372–378.

Dolbeer, R. A., S. E. Wright, J. Weller, and M. J. Begier. 2009. Wildlife strikes to civil aircraft in the United States 1990–2008. Serial Report 15. Federal Aviation Administration, Office of Airport Safety and Standards, Washington, D.C., USA.

Eastwood, E. 1967. Radar ornithology. Methuen, London, United Kingdom.

Federal Aviation Administration. 2010. Airport avian radar systems. Advisory Circular 150/5220-25. U.S. Department of Transportation, Washington, D.C., USA.

Felix, R. K., Jr., R. H. Diehl, and J. M. Ruth. 2008. Seasonal passerine migratory movements over the arid Southwest. Studies in Avian Biology 37:126–137.

Fischer, R. A., M. P. Guilfoyle, J. J. Valente, S. A. Gauthreaux Jr., C. G. Belser, D. Van Blaricom, J. W. Livingston, E. Cohen, and F. R. Moore. 2012. The identification of military installations as important migratory bird stopover sites and the development of bird migration forecast models: a radar ornithology approach. Project SI-1439. ERDC/EL TR-12-22. Strategic Environmental Research and Development Program, Alexandria, Virginia, USA.

Gauthreaux, S. A., Jr. 1972. Radar techniques for Air Force applications in avoidance of bird-aircraft collisions and improvement of flight safety. Interim Technical Report. U.S. Air Force Office of Scientific Research, Arlington, Virginia, USA.

Gauthreaux, S. A., Jr., and C. G. Belser. 1998. Displays of bird movements on the WSR-88D: patterns and quantification. Weather Forecasting 13:453–464.

Gauthreaux, S. A., Jr., and C. G. Belser. 1999a. Reply to "Displays of bird movements on the WSR-88D: patterns and quantification." Weather Forecasting 14:1041–1042.

Gauthreaux, S. A., Jr., and C. G. Belser. 1999b. Bird migration in the region of the Gulf of Mexico. Pages 1931–1947 in N. J. Adams and R. H. Slotow, editors. Proceedings of the 22nd International Ornithological Congress. BirdLife South Africa, Johannesburg, South Africa.

Gauthreaux, S. A., Jr., and C. G. Belser. 2003a. Bird movements on Doppler weather surveillance radar. Birding 35:616–628.

Gauthreaux, S. A., Jr., and C. G. Belser. 2003b. Radar ornithology and biological conservation. Auk 120:266–277.

Gauthreaux, S. A., Jr., C. G. Belser, and D. Van Blaricom. 2003. Using a network of WSR-88D weather surveillance radars to define patterns of bird migration at large spatial scales. Pages 335–346 in P. Berthold, E. Gwinner, and E. Sonnenschein, editors. Avian migration. Springer-Verlag, Berlin, Germany.

Gauthreaux, S. A., Jr., and J. W. Livingston. 2006. Monitoring bird migration with a fixed-beam radar and a thermal imaging camera. Journal of Field Ornithology 77:319–328.

Gauthreaux, S. A., Jr., J. W. Livingston, and C. G. Belser. 2008. Detection and discrimination of fauna in the aerosphere using Doppler weather surveillance radar. Integrative and Comparative Biology 48:12–23.

Griffin, D. R. 1972. Nocturnal bird migration in opaque clouds. Pages 169–188 in S. R. Galler, K. Schmidt-Koenig, G. J. Jacobs, and R. E. Belleville, editors. Animal orientation and navigation. NASA SP-262. National Aeronautics and Space Administration, Washington, D.C., USA.

Harmata, A. R., G. R. Leighty, and E. L. O'Neil. 2003. A vehicle-mounted radar for dual-purpose monitoring of birds. Wildlife Society Bulletin 31:882–886.

Harmata, A. R., K. M. Podruzny, J. R. Zelenak, and M. L. Morrison. 1999. Using marine surveillance radar to study bird movements and impact assessment. Wildlife Society Bulletin 27:44–52.

Horn, J. W., and T. H. Kunz. 2008. Analyzing NEXRAD Doppler radar images to assess nightly dispersal patterns and population trends in Brazilian free-tailed bats (Tadarida brasiliensis). Integrative and Comparative Biology 48:24–39.

Istok, M. J., A. D. Stern, R. E. Saffle, B. Bumgarner, B. R. Klein, N. Shen, Y. Song, Z. Wang, and W. M. Blanchard. 2008. Terminal Doppler weather radar for NWS operations: phase 3 update. Paper 6B.10. Proceedings of the 24th conference on interactive information and processing systems for meteorology, oceanography, and hydrology. American Meteorology Society, 20–24 January 2008, New Orleans, Louisiana, USA.

Kelly, T. A., R. Merritt, and G. W. Andrews. 2007. An advanced avian radar display for automated bird strike risk determination for airports and airfields (abstract). Proceedings of the 9th joint annual meeting. Bird Strike Committee–USA/Canada, 10–13 September 2007, Kingston, Ontario, Canada.

Kerlinger, P. 1982. The migration of common loons through eastern New York. Condor 84:97–100.

Klazura, G. E., and D. A. Imy. 1993. A description of the initial set of analysis products available from the NEXRAD WSR-88D System. Bulletin of the American Meteorological Society 74:1293–1311.

Klope, M. W., R. C. Beason, T. J. Nohara, and M. J. Begier. 2009. Role of near-miss bird strikes in assessing hazards. Human–Wildlife Conflicts 3:197–204.

Lack, D., and G. C. Varley. 1945. Detection of birds by radar. Nature 156:446.

Larkin, R. P. 1984. The potential of the NEXRAD radar system for warning of bird hazards. Pages 369–379 in Proceedings of the wildlife hazards to aircraft conference and workshop. Federal Aviation Administration, Charleston, South Carolina, USA.

Larkin, R. P., and L. Eisenberg. 1978. A method for automatically detecting birds on radar. Journal of Field Ornithology 49:172–181.

Larkin, R. P., and B. A. Frase. 1988. Circular paths of birds flying near a broadcasting tower in cloud. Journal of Comparative Psychology 102:90–93.

LGL Environmental Research Associates. 2000. National avian-wind power planning meeting III: proceedings. King City, Ontario, Canada. http://old.nationalwind.org/publications/wildlife/avian98/default.htm.

MacKinnon, B. 2006. Avian radar: demonstrated successes and emerging technologies. TP 8240. Airport Wildlife Management Bulletin No. 36. Transport Canada, Ottawa, Ontario, Canada.

McCracken, G. F. 1996. Bats aloft: a study of high-altitude feeding. Bats 14:7–10.

McCracken, G. F., and J. K. Westbrook. 2002. Bat patrol. National Geographic 201:114–123.

National Academy of Sciences. 2007. Appendix C: methods and metrics for wildlife studies. Pages 319–322 in Environmental impacts of wind-energy projects. National Academies Press, Washington, D.C.

Nohara, T. J. 2009. Could avian radar have prevented US Airways Flight 1549's bird strike? Proceedings of the Bird Strike North America conference. Bird Strike Committee–USA/Canada, 14–17 September 2009, Victoria, British Columbia, Canada.

Nohara, T. J., R. C. Beason, and P. Weber. 2011. Using radar cross-section to enhance situational awareness tools for airport avian radars. Human–Wildlife Interactions 5:210–217.

Nohara, T. J., P. Weber, A. Premji, C. Kranor, S. A. Gauthreaux Jr., M. Brand, and G. Key. 2005. Affordable avian radar surveillance systems for natural resource management and BASH applications. Pages 10–15 *in* IEEE 2005 international radar conference. IEEE, Washington, D.C., USA.

Plissner, J. H., T. J. Mabee, and B. A. Cooper. 2006. A radar and visual study of nocturnal bird and bat migration at the proposed Highland New Wind Development project, Virginia, fall 2005. Report to Highland New Wind Development. Harrisonburg, Virginia, USA.

Russell, K. R., and S. A. Gauthreaux Jr. 1998. Use of weather radar to characterize movements of roosting purple martins. Wildlife Society Bulletin 26:5–16.

Russell, K. R., D. S. Mizrahi, and S. A. Gauthreaux Jr. 1998. Large-scale mapping of purple martin pre-migratory roosts using WSR-88D weather surveillance radar. Journal of Field Ornithology 69:316–325.

Schmaljohann, H., F. Liechti, E. Bächler, T. Steuri, and B. Brudere. 2008. Quantification of bird migration by radar—a detection probability problem. Ibis 150:342–355.

Weber, P., T. J. Nohara, and S. Gauthreaux Jr. 2005. Affordable, real-time, 3-D avian radar networks for centralized North American bird advisory systems (abstract). Proceedings of the 7th joint annual meeting. Bird Strike Committee–USA/Canada, 14–18 August 2005, Vancouver, British Columbia, Canada.

Westbrook, J. K., and W. W. Wolf. 1998. Migratory flights of bollworms, *Helicoverpa zea* (Boddie), indicated by Doppler weather radar. Pages 354–355 *in* Proceedings of the second urban environment symposium and 13th conference on biometeorology and aerobiology, 2–6 November 1998, Albuquerque, New Mexico. American Meteorological Society, Boston, Massachusetts, USA.

Williams, T. C. 1984. How to use marine radar for bird watching. American Birds 38:982–983.

Williams, T. C., J. Settel, P. O'Mahoney, and J. M. Williams. 1972. An ornithological radar. American Birds 26:555–557.

Zakrajsek, E. J., and J. A. Bissonette. 2005. Ranking the risk of wildlife species hazardous to military aircraft. Wildlife Society Bulletin 33:258–264.

Zrnic, D., and A. Ryzhkov. 1998. Observations of insects and birds with a polarimetric radar. IEEE Transactions on Geoscience and Remote Sensing 36:661–668.

14 Avian Survey Methods for Use at Airports

Bradley F. Blackwell
Paige M. Schmidt
James A. Martin

Adverse effects and damage caused by interactions between humans and wildlife are increasing (De-Stephano and DeGraaf 2003). To manage wildlife effectively—whether to mitigate damage, to enhance safety, or to reach conservation goals—wildlife biologists must identify hazards posed by or to members of a particular species (i.e., a population) or guild, and then prioritize management goals and specific actions. We examine the special problem of managing birds to reduce hazards to aviation, particularly those species known to cause structural damage to aircraft when struck and that pose problems to airport facilities (Dolbeer et al. 2000, Cleary and Dolbeer 2005, DeVault et al. 2011). Effective management of hazardous species at airports requires knowledge of species abundance and how abundance varies over time. In this context, the quality of the sampling methodology used will influence a biologist's ability to accurately quantify avian hazards and to understand the ecological interactions of populations or guilds using airport environments.

Accurate quantification of avian hazards allows biologists to calculate the relative risk presented by each population or guild for a period and habitat, and relative to management actions. A hazard (whether a resource contributing to bird use or simply incidental use of the airport by a population or guild) represents a particular state or condition within the airport environment that can affect the probability of bird strikes. In contrast, we define risk as the relative conditional probability of damage to an aircraft posed by a species, if struck, and the probability of the strike occurring (Schafer et al. 2007,

Blackwell et al. 2009). Avian survey data form the foundation for identifying management priorities, reducing risks associated with avian hazards to aviation safety, and evaluating the effectiveness of management actions.

Defensible data collection, analysis, and accurate findings are imperative to justify management options to other agencies and, increasingly, to a critical public (Anderson 2001). Lethal control of birds, although regulated, is an integral component of wildlife hazard reduction to mitigate strike risk at airports (see Cleary and Dolbeer 2005; Chapter 7). Despite this importance, public support for lethal control measures in wildlife management, regardless of the justification for their use, is declining. As a result, increased documentation is required to receive necessary permits, and there is need to directly demonstrate the efficacy of lethal control measures when used (Blackwell et al. 2002, 2009; Engeman et al. 2009; Runge et al. 2009). However, the union between direct management, particularly lethal control, and scientifically rigorous data collection has proven useful for demonstrating and justifying lethal control for endangered species recovery (Engeman et al. 2005, 2009), as well as for enhancing aviation safety (e.g., Dolbeer et al. 1993, Seamans et al. 2009). A demonstration of scientifically sound methods in the collection of survey data is increasingly necessary to justify and legally defend various management actions—particularly lethal control—even in situations involving human health and safety (Messmer et al. 1997, Reiter et al. 1999, Conover 2002).

Despite the need for scientific rigor, resource limitations often require that biologists base management decisions on brief samples or "snapshots" of target populations. The process used to take these snapshots, if based on sound sampling theory, will yield accurate inference as to population abundance or trends, habitat influences, seasonal dynamics, and response to management actions (Morrison et al. 2008). As outlined by Cochran (1977) and adapted here for application to the airport environment, the sample survey should be based on six primary steps: (1) define the objective, (2) delineate the target population, (3) determine the data necessary to address the objective, (4) identify and correct for factors that influence accuracy of the estimate, (5) select appropriate methods of measurement, and (6) select appropriate data management and analysis procedures. The survey objective will dictate aspects of the subsequent steps, as will available resources.

In this chapter we use published sampling theory and methods to provide airport biologists with (1) the means to design and implement an avian survey at an airport that will maximize accuracy in quantifying avian hazards; (2) an understanding of bias and precision, and their influences on the quantification of avian hazards; (3) suggestions on how to quantify avian hazards and how to use these data to estimate relative risk to aviation safety posed by a particular species or guild by time period and habitat type; and (4) knowledge of how data can be used to prioritize management goals. Our recommendations are intended to compliment Federal Aviation Administration (FAA) procedures for Wildlife Hazard Assessments (WHAs) and subsequent management at airports (Cleary and Dolbeer 2005).

Define the Objective

Defining objectives for a wildlife study or assessment is the first step in the process of designing and implementing the effort. Clearly defined objectives allow biologists to delineate target populations, to collect representative data using an appropriate survey method, to manage data, and to identify appropriate analysis methods. In the context of avian surveys for hazard assessment purposes, the regulations that require the assessment often help define study objectives.

The FAA (2004a) dictates that a certificated airport must take immediate action to alleviate wildlife hazards whenever they are detected, and must ensure that a WHA is conducted when specified criteria relating di-

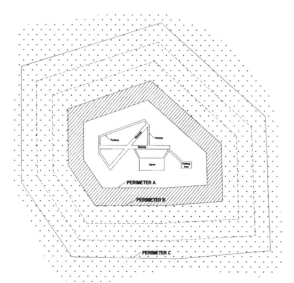

Fig. 14.1. Federal Aviation Administration (FAA) siting criteria outlining separation distances within which airports must manage attractants to hazardous wildlife. Perimeter A: for airports serving piston-powered aircraft, hazardous wildlife attractants must be 1,524 m (5,000 feet) from the nearest air operations area. Perimeter B: for airports serving turbine-powered aircraft, hazardous wildlife attractants must be 3,048 m (10,000 feet) from the nearest air operations area. Perimeter C: 8-km (5-mile) range to protect approach, departure, and circling airspace. From FAA (2007)

rectly to wildlife strikes or the potential thereof exist at the airport. The WHA must include the "identification of the wildlife species observed and their numbers, locations, local movements, and daily and seasonal occurrences" (FAA 2004a). Under this regulation, the broad objective of a survey is to identify and quantify wildlife hazards on and near airport properties, and the implication is that a management protocol (i.e., a wildlife hazard management plan; Cleary and Dolbeer 2005) will be implemented to reduce or remove the identified hazards. Airport properties include the air operations area (AOA), defined as the space designated for takeoff, landing, and surface maneuvers of aircraft (see FAA 2004a). However, wildlife attractants might also reside in areas defined by FAA siting criteria for certificated airports (i.e., within 1.5 km [1 mile] of a runway for airports servicing piston-powered aircraft only and within 3.0 km [2 miles] of a runway for airports servicing turbine-powered aircraft; FAA 2004b, 2007; Fig. 14.1).

Fig. 14.2. Runway protection zones at Seattle–Tacoma International Airport, Seattle, Washington, USA. From Schafer et al. (2007)

A further implication of this broad objective is that airport biologists can, on the basis of these survey data, prioritize management goals. To this end, the survey data should allow an assessment based on calculation of risk or the probability of a damage-causing wildlife strike. The working objectives for the survey might be: (1) to quantify seasonal abundance of a population or guild within specific airport habitats or habitats immediately bordering airport property; (2) to calculate the relative risk posed by a population or guild by season, habitat type (e.g., local attractants), or airport environment (e.g., AOA, runway protection zone; Fig. 14.2); (3) to use estimates of relative risk to justify management recommendations at and near the airport; and (4) to quantify the effect of management actions on subsequent estimates of strike risk. Inherent to the survey objective are means to demonstrate risk, to enact sound wildlife management on the airport, to establish a defensible foundation for working with property owners and municipalities within the airport siting criteria, and to evaluate the effect of wildlife hazard reduction and risk mitigation.

Defining Target Populations

The target population is the population about which information is required (Cochran 1977, Morrison et al. 2008). Although numerous wildlife species are hazardous to aviation, in this chapter we focus only on avian hazards (Dolbeer and Wright 2009). Both diurnal and nocturnal bird species pose hazards to aviation, but the survey methods that apply to nocturnal species, particularly sampling equipment (e.g., forward-looking infrared cameras or avian radar systems; Chapter 13) and associated constraints, are beyond the scope of this chapter. Our focus is on quantifying use of airport habitats by diurnal bird species.

In the context of a WHA, and in reference to FAA (2004a), airport biologists should anticipate gathering data on multiple avian species during a survey. The initial site visit will provide anecdotal information on species using the airport, as well as potential attractants (Cleary and Dolbeer 2005). Avian species that appear frequently in an airport's strike records or database like that maintained by the FAA, particularly those species involved in strikes resulting in substantial damage (Dolbeer et al. 2000, 2010), will be a primary focus for airport biologists. Dolbeer (2006) found that of those bird strikes occurring at ≤152 m (500 feet) above ground level (AGL), Passeriformes, gulls and terns (Laridae), doves and pigeons (Columbidae), and raptors were the guilds most frequently struck. For strikes at >152 m AGL, waterfowl (Anatidae), gulls and terns, passerines, and vultures (Cathartidae) were the most frequently struck. In addition to assessing strike hazards, it is conceivable that airport biologists could be called upon to make management recommendations for species occupying habitats outside the AOA, including those deemed nonhazardous to aviation, of particular conservation concern (e.g., state or federally protected species, grassland bird species; Blackwell et al. 2009), or species of concern that pose a direct strike hazard.

Necessary Data

Biologists conducting assessments of bird communities at airports must predetermine the data necessary to address the identified objectives. If this step is ignored, one might collect unnecessary data, wasting time and resources at the expense of data necessary to meet the objectives of quantifying avian hazards. Driving the perimeter of the airfield twice monthly can provide perspective on birds attracted to roads and edge habitats and identify other animal attractants at the airport, but this approach will never yield accurate data on population abundance within those habitats, or similar data for populations with more specific habitat requirements. Identification of the data necessary to address the specific objectives of the assessment will aid survey design and conduct, as well as data management and analysis.

Data collected by airport biologists generally comprise naive counts (i.e., counts that are uncorrected for imperfect detection; MacKenzie et al. 2002) of individual birds and flocks, including numbers of individuals within the flocks, identified to species or guild. These data are collected during avian surveys at airports using a point-transect approach (Cleary and Dolbeer 2005) that parallels the North American Breeding Bird Survey (BBS; Sauer et al. 2008). Point transects and variations thereof (e.g., Emlen 1977, Reynolds et al. 1980, Bollinger et al. 1988, Bajema et al. 2001) can offer coverage of a breadth of habitats, minimize observer effect on avian behavior (e.g., potentially "pushing" birds ahead of the observer during a transect survey), and sample within fixed areas. In the context of an airport, naive counts made via point transect or comparable methods are also an effective means of identifying habitats and land uses that potentially serve as attractants to birds that pose strike hazards to aircraft (Cleary and Dolbeer 2005). But naive count data do not allow for accurate inference of relative abundance of particular populations (i.e., one cannot rank relative hazards) unless the methods used to obtain these counts are standardized for the conditions under which they are measured (Caughley 1977).

Sampling efforts should be tied to space and time (e.g., Buckland 2006) and adjusted for biases (Link and Sauer 1998, Runge et al. 2009), particularly imperfect detection (Lynch 1995, MacKenzie et al. 2002, MacKenzie 2005). Otherwise, the count data (e.g., BBS data) can be ecologically ambiguous. Specifically, naive counts cannot be associated with a probability distribution, which is integral to assessing the accuracy and variability in an estimate of population abundance and, by extension, standardizing how management priorities are determined (see below).

Imperfect detection is essentially the inability to detect or correctly identify birds that are present (e.g., Lynch 1995, MacKenzie et al. 2002), or recording birds as detected when they are not actually present. As a result, this error or bias is introduced into the data analysis. Bias in data collection is considered a systematic error that can result in under- or overestimation of the parameter of interest, such as population abundance (Thompson 2002). Error in estimates of population abundance can subsequently influence estimates of relative risk and the prioritization of management efforts. Birds that use open areas (e.g., eastern meadowlarks [*Sturnella magna*]) might be more easily detected than species that use wooded areas (e.g., wild turkey [*Meleagris gallopavo*]), possibly resulting in higher counts for species or individuals preferring open areas (Ellingson and Lukacs 2003). Bias introduced by variability in detection due to habitat utilization might lead biologists to conclude that hazardous birds use open areas more often, or that populations using open areas are more numerous and pose a greater risk than those that use wooded areas, when the opposite could be true.

Many species hazardous to aviation are readily detectable, such as the European starling (*Sturnus vulgaris*) and rock pigeon (*Columba livia*). Still, errors can occur in estimating flock size and composition, even for the most obvious species. Factors that can influence relative numbers of observed individuals include temporal variation in flocking behavior (e.g., during breeding season), variation in individual behavior, season (e.g., leaf off versus leaf on), and response to recent management actions (Ellingson and Lukacs 2003). We caution that, without means of correcting for bias associated with imperfect detection, data obtained from avian point-transect counts will yield only an index count and cannot be used reliably to estimate risk. We discuss means of estimating detection bias in Methods of Measurement below.

Factors That Influence Accuracy

Biologists use a sample to estimate site abundance with regard to avian hazards at airports and to determine how various factors (e.g., habitat, season, detection, management) might influence those estimates and, ultimately, relative risk. One must understand what influences the quality of observation data (e.g.,

Thompson 2002, Morrison et al. 2008). First, the estimated parameter (e.g., abundance) should be unbiased or close to the true value. Second, the estimate should be precise, whereby its value fluctuates minimally over repeated samples within an ecologically important period. Both the bias and precision associated with the collected survey data will determine the validity of the estimate.

We discussed bias due to imperfect detection in the preceding section, but other factors can introduce bias or affect the variability of the survey data (i.e., precision) and possibly accuracy. For example, bird counts are affected by observer ability, observer behavior during the survey (i.e., birds attracted to or repelled by the observer), season, time of day, temperature, wind, precipitation, cloud cover, and light intensity (Rosenstock et al. 2002, Thompson 2002). In addition, the presence of predators or other disturbances, including harassment, will affect bird behavior and variability in counts. Indices of relative abundance (e.g., naive count data), which are routinely used in WHAs, can be both precise and inaccurate due to consistent bias or consistent sampling at wrong times. Subsequently, the potential for inconsistency in bias also precludes comparison of indices of relative abundance, as these data do not provide information on how bias influences the proportion of the true, undetected value (Bart and Earnst 2002). From an ecological standpoint, we will likely never know the true value for a parameter at any given time (Burnham and Anderson 1998). However, we can approximate "ecological truth" by collecting data in a manner that allows adjustment for potential biases and that minimizes variability (i.e., increases precision) in estimates of population abundance and density within habitats or time periods, whether during an ecological season or predefined period (Thompson 2002).

Bias associated with factors that can influence accuracy of survey data and that are outside the biologist's control must be reduced through careful standardization of survey methods, as noted in guidelines for avian surveys at airports (Cleary and Dolbeer 2005). Spatial and temporal distribution of the survey effort can influence the data collected, and these factors are within the control of the biologist via careful sample design (see below). Differential availability and use of a habitat by a given population or guild can introduce variability and bias into estimates of population abundance if not accounted for in the allocation of survey effort. Compounding such biases are temporal variation in habitat availability (e.g., winter or summer, wet or dry), period of use (e.g., migration, breeding season), and variation in daily activity of species. Some species (e.g., vultures) increase their activity later in the day as thermals increase; failure to sample these populations during periods that correspond to peak activity will result in bias (Stolen 2000, Runge et al. 2009). Differences in survey data between habitats can be more of a reflection of the distribution of survey effort between habitats (including associated influences such as distance to another habitat type) or sampling time than actual differences in abundance.

We can reduce bias associated with factors outside the biologist's control with careful standardization of survey methods. Biologists can also reduce the bias of factors within their control through careful sampling design and allocation of effort to incorporate both spatial and temporal variability in the airport environment. Such efforts will improve accuracy in the quantification of use of airport and near-airport habitats by bird populations.

Methods of Measurement

After following the preceding steps, biologists will have identified the objectives of the survey, the target populations, the necessary data to be collected, and the factors that influence the accuracy of hazard quantification and subsequent calculation of risk. The design of an avian survey also requires that biologists consider the total survey effort necessary to meet objectives as well as the allocation of the survey effort in both space and time. First, the survey should adequately sample the habitats at the airport (i.e., up to 3 km [1.9 miles] from a runway edge; Fig. 14.1) and its potential attractants. We suggest that airport biologists use a geographic information system (GIS) to systematically locate observation points for the survey, spanning the airport environment, including terminal buildings and large rooftop areas. These points represent centroids of cells whose areas correspond to the estimated sighting distance for the least detectable (e.g., because of habitat use or behavior) species of concern with regard to aviation safety. In this systematic layout of sampling points, two centroids will be separated by twice the predetermined sighting distance. In addition, the complexity of airport habitats and the total area of interest determine the total number of cells (i.e., general aviation airports usually require fewer points than large, Part 139–certificated airports). The goal, however, is to systematically "cover" airport habitats and abutting properties with cells identical in area.

Given the "population" of cells across the airport, several options are available to sample these cells. If habitats at the airport are represented disproportionately by areas maintained for aesthetics (e.g., wetlands, natural grasslands, or in forest), biologists might consider stratification (e.g., see Buckland et al. 2001), an approach by which cells within predefined habitats are selected for survey relative to their proportionate representation of total airport area. However, we recommend broad classifications of habitat type (e.g., rooftop, managed grassland, runway, wetland, etc.) to avoid issues with inadequate sample size. Also, because habitats at an airport might change due to development, mitigation, or management recommendations from airport biologists, stratifications might also change.

The simplest approach to provide a representative sample of airport habitats—one that does not necessitate a redesign of the sampling approach as habitats change—is a basic random sample of cells delineated across the airport (as described above). Under this approach, biologists will randomly select a total of 20 cells (or as many as possible up to 20, depending upon airport size). These 20 cells will be used for each of three daily observation periods—morning (30 min before sunrise to 1000 hours), midday (1200 to 1500 hours), and evening (1600 hours to 30 min after sunset)—that would be conducted during each season or period of interest. A survey protocol involving 20 cells allows biologists to account for variance in encounter rates and for constructing confidence intervals about mean encounter rates (e.g., Buckland et al. 2004). Survey data might reveal a sudden increase in numbers of a particular population, but whether these data reflect a pulse of birds moving through the airport or a consistent pattern of use can be discerned only through adequate survey coverage of airport habitats and frequency (see below).

Because bird movements within the airport environment vary by season (e.g., breeding periods versus migration), and because it is crucial to avoid the bias

of "pushing" birds ahead of the survey (see Buckland et al. 2001, 2004), we recommend that biologists restrict the 20 cells to those with separations of at least 500 m (1,640 feet) between centroids (depending upon minimum sighting distance and therefore cell radius). This restriction necessitates a systematic examination of the location of the 20 cells and adjustment for the distance between cells. Because the initial set of 20 cells is selected randomly, we do not foresee issues with bias due to the adjustment for cell intervals. In addition, a replacement for a cell that has restricted access (e.g., whether an official or a logistical constraint) or site conditions that prevent adequate sighting of birds (e.g., a point falling within a mature corn stand) should also be selected at random and with regard to cell radius.

We base our suggestion of 20 cells on the necessity of adequate coverage of airport habitats and the constraints of time allocation. If we assume a minimum sighting distance of 200 m (656 feet; representing the radius of a cell), a random sample of 20 cells comprises 251 ha of airport and abutting properties that are surveyed. We note that the average area for a certificated airport located in the contiguous USA is 761 ha (DeVault et al. 2012). Biologists might choose to randomly select 20 cells for observations during each season. A season represents an ecologically significant time period with respect to species typically observed at the airport or those anticipated to move through the region. Whether the same 20 cells are surveyed across seasons or sets of 20 are selected randomly for each season, we suggest that surveys proceed through a full calendar year, allowing comparison of population or guild abundance estimates across seasons.

Biologists are interested in discerning how a population is represented in various airport habitats over time, thus justifying a longer period of surveys. For most populations, airport habitats likely represent a small portion of the overall range during a given season (there will be exceptions, such as with rock pigeons; see Martin et al. 2011). The presence of members of a species in airport habitats can be considered as random, and the associated survey data can be interpreted as "use" as opposed to "occupancy" (MacKenzie 2005; see also Occupancy Models, below).

The start time per survey period and starting location will also be randomized. We assume a 3-min observation period, meaning a subset of 20 cells can likely be surveyed, considering travel time, within 2 hr. As for survey effort within season, we base our recommendation on that of MacKenzie and Royle (2005), who suggest that sampling units (cells) be surveyed a minimum of three times within a season when detection probability for a species is >50% per survey. In an airport environment, considering that data from multiple species with different detection probabilities will be obtained, we recommend that biologists plan for a minimum of three surveys per month.

After designing and allocating the survey, biologists collect data in the field. In practice, biologists start at the first cell (randomly selected from the sample of 20 cells), prepare binoculars and data sheet, spend 3 min observing the area around the centroid up to the maximum predefined radius, and record any birds that are seen. We do not recommend using aural detection at airports (i.e., identifying birds by song or call), because noise interference inevitably affects detection of sound intensity (energy content of the call or song), pitch (song frequency), or modulation (variation in pitch or intensity; see Alldredge et al. 2007, and citations therein). If a bird or flock is first detected aurally and confirmed visually to be within the cell bounds, however, that observation should be recorded. The biologists should also consider recording an activity code (e.g., loafing or foraging) for the observation, so as to inform potential hazard management decisions within the area of the cell.

The 3-min observation period minimizes the potential evasive movements and avoidance of the area by birds and attraction to the observer (e.g., some members of Corvidae; Scott and Ramsey 1981; see also Rosenstock et al. 2002). The observer should then move directly to the next preselected cell (based on proximity), maintaining as best as possible a consistent time interval between cells. We stress that surveys should never be conducted from within a vehicle (as per Cleary and Dolbeer 2005), as doing so inhibits visibility of the entire cell, thus increasing bias due to imperfect detection.

When biologists record observations from a specific centroid, it is assumed that the birds are associated with the cell bounding that observation point (i.e., observations of birds outside the immediate bounds of the cell are recorded as incidental). But not every bird

or flock observed will be on the ground within the cell. For birds that are flying and deemed to be using the habitat within the cell, assuming a vertical extension of the bounds of the cell (e.g., birds entering the cell volume to land; raptors hovering over prey), the observations are recorded as if the birds were on the ground. If possible, the observer should also estimate the birds' altitude using the height of features at the airport (e.g., the control tower; Hoover and Morrison 2005) as reference points. We acknowledge that such estimates might be possible only for flocks entering the cell at relatively low altitudes (e.g., ≤30.5 m [100 feet]). Soaring raptors and vultures often fly at altitudes that are impossible to associate with a specific airport habitat; such observations should, however, be recorded as incidental to the primary survey data. Although altitude estimates are not components of population abundance estimates, these data can prove useful with regard to enhancing the spatial component of risk assessments.

As noted above, biologists generally record individual birds and flocks, including numbers of individuals within the flocks, as part of a species or guild. In some analytical approaches, particularly the Program Distance approach (Buckland et al. 2001; see below), analyses are based on either individuals or clusters (e.g., flocks). Animals behaving in groups, such as flocks, cannot be considered independent observations in subsequent analyses. When observing flocks, biologists record the number of birds within a flock only if the flock center lies within the cell area; if some individuals of the flock lie within the cell area but the flock's center lies beyond it, biologists record these data as incidental, and they should not be used in analyses (see Buckland et al. 2001). Birds noted during travel between points also should be recorded separately as incidental (Hutto et al. 1986).

In the event that biologists must react to a hazardous situation during the survey, including the need to disperse birds in the path of an approaching aircraft, data for the cells close enough for the birds to be affected by the disturbance should be noted as "missing," and the reason should be stated (see example in Table 14.1). Importantly, biologists should not record "zero birds" due to dispersal activities, as a "zero" represents actual data and has bearing on population or guild abundance estimates. Any increase in missing data due

Table 14.1. Sample data sheet for conducting an airport avian survey.

Date	Time	Cell	Habitat	Species[a]	Cluster size[b]	Comments
1 September 2013	1622	1	–	–	0	No birds.
1 September 2013	1630	2	Grass	EUST	20	Multiple flocks in area.
1 September 2013	1630	2	Grass	EUST	10	After count, dispersed flock as aircraft approached.
1 September 2013	1640	3	Grass	EUST	(10)[c]	Same flock as dispersed from cell 2; do not use for density.
1 September 2013	1650	4	–	Gull	Missing	Dispersal action necessary before count.

[a]Indicate the species by using a four-letter code for the common name (e.g., EUST for European starling), or by listing the guild category (e.g., blackbirds, starlings) if the bird species is not identifiable.

[b]List the number of individual birds observed as the cluster size. Such grouping data are used in analyses of density (see Buckland et al. 1993).

[c]Identify observations that might confound estimates of density, such as birds moving in response to the observer (noted here by parentheses).

to hazard mitigation should be offset by an increase in sampling effort.

Data Management and Analysis

Even when the objectives and target populations are clearly defined and suitable data with required accuracy are collected using appropriate methods, spurious conclusions and recommendations are possible if improper data management and analysis procedures are used. Survey data should be recorded to a spreadsheet or database as soon as possible following a survey (Table 14.1). We suggest that observations also include the appropriate family category (i.e., American Ornithologists' Union classifications) or a guild category reflecting birds documented as hazardous to aircraft (Dolbeer et al. 2010; Table 14.1). Each line of data for an observation will include the cell number, flock size (i.e., the number of individuals within the flock), population or guild, survey time, and date. These raw survey data are then available for a basic descriptive analysis that reflects an index of abundance (i.e., the total number of detections or frequency of detections) unadjusted for error (Burnham 1981; Buckland et al. 1993; Anderson 2001, 2003; see also Rosenstock et al. 2002). Observers can calculate the index for each population or guild by period and habitat, or both. Again, we caution that formal conclusions about relative habitat use by different populations or guilds or about relative abundance should not be based on raw or naive count data alone. We agree with Burnham (1981) that using the count of birds per unit effort as an index of abundance does not provide a scientifically sound or reliable estimate of abundance.

Several analytical options are available that incorporate detection histories, given attention to potential biases in survey design and conduct, and the assumptions associated with the particular analysis used to estimate population abundance (e.g., distance sampling; Buckland et al. 1993, 2001, 2004; modeling based on the relationship between detection probability and abundance distributions; Royle and Nichols 2003). These methods allow biologists to build on information gleaned from naive count data obtained from a well-designed survey to discern patterns of use relative to probability distributions. The Double Sampling approach for estimating population density, advocated and described in detail by Bart and Earnst (2002), gives density (D) as the number of individuals (N) observed per unit area (A; or $D = N/A$), if we assume all animals are detected. Because detection is rarely perfect, however, biologists must correct the number of observed individuals to account for missed detections in order to produce an unbiased estimate of density. Density estimation is a departure from the common practice of using naive WHA counts at airports (Cleary and Dolbeer 2005, Schafer et al. 2007).

Under the Double Sampling method, biologists use the sampling approach described above as an initial "rapid" survey. In addition, they choose six of the 20 randomly selected cells (described in Methods of Measurement, above) for an intensive survey to be conducted soon after the previous rapid survey. The intensive survey entails a systematic "walk-through" of the fixed-radius cell, noting all birds or flocks observed in the cell (as described above) or flushed from within the cell. The intensive survey data represent the actual number of birds using the cell at that time. The estimate of density (D) is obtained as per Bart and Earnst (2002):

$$D = (\overline{x}') / (\overline{x} / \overline{y}),$$

where $\overline{x}'$ represents the mean number of birds or flocks of a particular population or guild recorded per cell during the rapid survey; $\overline{x}$ is the mean number of birds recorded per cell across the subsample of six cells during the rapid survey; and $\overline{y}$ is the mean number of birds actually present per cell across the subsample of six cells (i.e., counted during the intensive survey). This approach works best when results from the rapid survey are highly correlated with actual density. Specifically, if $\overline{y}$ is biased, then D will also be biased. We recommend that the intensive survey be conducted immediately following the rapid survey.

The ratio of the mean count per cell in the subsample obtained during the rapid survey to the actual mean density as determined via the intensive survey of the cells in the subsample is used to adjust the results from the rapid survey. Bart and Earnst (2002) provide further detail about estimating standard errors about D, precision of the index ratio ($\overline{x} / \overline{y}$) in the subsample, and incorporating cost estimates. Although Bart and Earnst (2002) note that for their study the surveyors conducting rapid surveys of plots included in the intensive surveys had no prior experience with the plot, such a division of duties is not logistically feasible due to the constraints associated with staffing biologists at airports. Further, the authors focused on nest detection, whereas airport biologists obtain count data of individuals and flocks within the cell. Despite these differences in the Bart and Earnst (2002) field protocol and conditions found at most airports, we contend that the Double Sampling method would enhance the accuracy of density estimates for bird populations or guilds using airport environments.

Alternative Approaches

Distance Sampling

Distance sampling uses the distance from the observation point to an individual bird or flock to estimate a detection probability, which is then used (in the Distance software package) to calculate density (Buckland et al. 1993, 2001, 2004). Distance estimates for each

bird or flock are collected at the time of observation. The collection of additional covariates such as habitat variables allows for the calculation of more accurate detection probabilities and in turn more accurate density estimates. Covariates may be collected at the time of observation or at a later date using GIS or other stored data sets. Distance sampling requires few extra resources when compared to naive counts, but relies on several assumptions:

1. Objects at the line or point are detected with certainty. This assumption should be achievable in the airport environment, except under special circumstances such as species emitting calls only out of the observer's line of sight when noise interference is high.
2. Objects do not move in response to the observer or before detection, an assumption that can be met with proper field protocols such as undisruptive movements to and from survey locations. Additionally, if survey periods are kept short (e.g., ≤5 min), birds are not likely to move.
3. Distance measurements are exact. With the aid of laser range finders and given that most detections at airports are visual, this assumption is achievable. Furthermore, if cells have a small area, distances will be truncated to reduce bias. We recommend taking a spot-mapping approach, where bird locations are placed on an aerial image (or a map produced via the GIS; noted above) and relative to the transect or point to aid in distance calculations.

As noted above with regard to birds aerially foraging over sampled habitat, one must assume a vertical extension from the bird to a point on or off the line (noted by the observer relative to a particular landscape feature). The distance to that point from the line is then measured as described above.

Distance sampling can be a robust approach to estimating abundance or density for birds. In general, however, >60 observations are needed for each population or guild to gain reliable density estimates (Buckland et al. 2001). Alldredge et al. (2007) provide an applicable approach for combining multiple populations or guilds into a common framework to produce more reliable estimates for those groups lacking sufficient data to be modeled alone. This approach holds promise if distance sampling is used in the airport environment.

Mark-Recapture Approaches and Extensions

Mark-recapture methods using marked animals (e.g., bands) are not feasible for airport monitoring, but extensions to the mark-recapture framework that involve indirect "marking" and "recapturing" are feasible. These are based on repeated or replicated observations and are used to estimate detection probabilities. Multiple observer methods make use of two or more observers working either independently or collaboratively to account for individuals missed by each observer (Nichols et al. 2000). Removal models delineate the survey into distinct time periods (e.g., 0–3, 4–5 min) and use detections (i.e., captures) within time periods to develop a capture history across the entire sampling unit (Farnsworth et al. 2002). Time-to-detection methods use multiple, discrete vocalizations of individuals to develop a detection probability within a mark-recapture framework (Alldredge et al. 2007). There are many combination methods that attempt to further refine estimation of density, as well (e.g., double-observer distance sampling). The removal and time-to-detection methods are advantageous when estimating availability probabilities is needed (Diefenbach et al. 2007). Such cases are likely rare, considering again that we suggest visual detections in the airport environment. For species of conservation concern, such as some grassland birds, collecting data in a manner consistent with removal and time-to-detection models would be prudent, especially considering that this approach requires very little extra effort.

Occupancy Models

Occupancy sampling and modeling is an approach that uses repeated (more than two) observations of sites (e.g., cells) to estimate the state parameter (e.g., probability of occupancy, abundance) and the observation process (i.e., detection probability; MacKenzie et al. 2002). The simplest form of occupancy sampling is presence/absence data; however, these methods have been extended to model abundance (Royle and Nichols 2003). These methods can also be used to model

resource use depending on objectives and the assumptions of the sampling endeavor. The main advantage of these general classes of models is their overall flexible utility and intuitive interpretation. The major disadvantage is the necessity to repeat surveys, which increases effort. Not all surveys must be repeated, however. For example, if 20 cells are to be sampled, only 10 might need to be repeated. Additionally, stopping rules can be initiated when animals are detected, further reducing effort. The number of repeated surveys is negatively correlated with detection probability and probability of occupancy. Mackenzie and Royle (2005) provide a thorough treatise on survey allocation and design. In general, for highly detectable species, two to three survey occasions are needed for reliable estimates per season. If multiple ecologically relevant seasons exist (e.g., breeding and nonbreeding), it will be necessary to sample within each season.

Strike Risk

Once estimates of population or guild densities relative to habitat (e.g., short grass) and time period (e.g., breeding season, migration) have been obtained, biologists can more accurately quantify relative hazard. But we contend that quantification of hazards alone is inferior as a means of prioritizing management goals, because it does not account for the likelihood that a hazardous bird will be struck or for the damage caused by that strike, and one should not assume local population density to be correlated directly and positively with the probability of a bird being struck. In most cases we would not expect snow geese (*Anser caerulescense*) to be as locally dense as savannah sparrows (*Passerculus sandwichensis*), yet between 1990 and 2007, 68 strikes of snow geese were reported to the FAA (Dolbeer et al. 2010). Of those strikes, 78% caused damage, 38% had a negative effect on flight, and 54% involved strikes of more than one animal (Dolbeer and Wright 2009). Based on these data, snow geese were ranked as the third most hazardous wildlife species struck by aircraft and the most hazardous bird species struck. In contrast, for the same time period, 68 strikes of savannah sparrows were reported; of those strikes, 1% caused damage, 0% had a negative effect on flight, and only 7% involved strikes of more than one animal (see also DeVault et al. 2011).

We suggest that effective prioritization of population management at airports entails an assessment of the risk of damage from wildlife strikes (see Schafer et al. 2007). In this context, a risk assessment would reflect an estimate of a population's frequency of occurrence within critical locations at and near the airport (see also Martin et al. 2011) and associated strike damage metrics. A risk assessment has the following components (Graham et al. 1991): (1) a conceptual understanding of the sources of the problem (e.g., habitat attractive to hazardous wildlife at and near the airport); (2) realistic end points or potential events (e.g., a hull loss; Dolbeer et al. 2000, 2010); (3) mechanisms by which the sources contribute to the defined end points (e.g., is substantial strike damage related to a particular aircraft type or species struck at the airport?); (4) a spatiotemporal estimate of exposure to the problem sources (population or guild density data by habitat and time obtained via the avian survey); and (5) a quantification of potential effects (i.e., the calculation of risk based on components 1–4). Again, in the context of airports, seasonal demographic cycles of populations using particular habitats (e.g., agriculture near an airport) should be evaluated relative to population density estimates within critical airspace to better discern the contribution of habitat to bird-strike risk (Baxter and Robinson 2007). The bird-strike risk assessment should include, at a minimum, population or guild density estimates from the survey and associated strike statistics for those guilds, such as strike frequency for the specific airport and associated damage or damage statistics from the FAA (see Blackwell et al. 2009, DeVault et al. 2011). Other components might include data on aircraft types serviced by the airport and number of aircraft movements relative to seasonal abundance estimates of hazardous populations, as well as spatiotemporal associations of populations (Martin et al. 2011) and aircraft movements relative to altitude (J. Belant and J. Martin, unpublished data). In its most basic format (i.e., without incorporation of concurrent data on aircraft movements and spatiotemporal aspects of bird use of the AOA), however, risk can be expressed as the product of the relative frequency of each guild (i.e., its seasonal density estimate) and the proportion of bird strikes involving the guild that have resulted in damage to aircraft (across U.S. civil airports and civil aircraft).

Summary

We have purposely focused our recommendations on the quantitative aspects associated with design and conduct of an avian survey, with unique application to the airport environment. We have stressed the need for survey data to be ecologically relevant and accurate, such that management guidelines are based on defensible data. However, "real world" issues—regulatory, labor, and financial constraints, as well as the dynamics of airport environments—will inevitably influence survey methods. Though we do not advocate the use of naive count data in estimating relative abundance or habitat use, for example, we recognize that animal observations obtained by airport biologists outside of a standardized sampling protocol are important for identifying potential hazards to aviation safety. We recommend developing training materials for airport biologists that incorporate information provided in this chapter relative to constraints affecting survey design and conduct, so as to move effectively from concept to practice.

LITERATURE CITED

Alldredge, M. W., K. H. Pollock, T. R. Simons, and S. A. Shriner. 2007. Multiple-species analysis of point count data: a more parsimonious modeling framework. Journal of Applied Ecology 44:281–290.

Anderson, D. R. 2001. The need to get the basics right in wildlife field studies. Wildlife Society Bulletin 29:1294–1297.

Anderson, D. R. 2003. Response to Engeman: index values rarely constitute reliable information. Wildlife Society Bulletin 31:288–291.

Bajema, R. A., T. L. DeVault, P. E. Scott, and S. L. Lima. 2001. Reclaimed coal mine grasslands and their significance for Henslow's sparrows in the American Midwest. Auk 118:422–431.

Bart, J., and S. Earnst. 2002. Double sampling to estimate density and population trends in birds. Auk 119:36–45.

Baxter, A. T., and A. P. Robinson. 2007. Monitoring and influencing feral Canada goose (*Branta canadensis*) behaviour to reduce birdstrike risks to aircraft. International Journal of Pest Management 53:341–346.

Blackwell, B. F., T. L. DeVault, E. Fernández-Juricic, and R. A. Dolbeer. 2009. Wildlife collisions with aircraft: a missing component of land-use planning for airports. Landscape and Urban Planning 93:1–9.

Blackwell, B. F., M. A. Stapanian, and D. V. Weseloh. 2002. Evaluating dynamics of the double-crested cormorant population on Lake Ontario. Wildlife Society Bulletin 30:345–353.

Bollinger, E. K., T. A. Gavin, and D. C. McIntyre. 1988. Comparison of transects and circular-plots for estimating bobolink densities. Journal of Wildlife Management 52:777–786.

Buckland, S. T. 2006. Point-transect surveys for songbirds: robust methodologies. Auk 123:345–357.

Buckland, S. T., D. R. Anderson, K. P. Burnham, and J. L. Laake. 1993. Distance sampling: estimating abundance of biological populations. Chapman and Hall, London, United Kingdom.

Buckland, S. T., D. R. Anderson, K. P. Burnham, J. L. Laake, D. L. Borchers, and L. Thomas. 2001. An introduction to distance sampling: estimating abundance of biological populations. Oxford University Press, New York, New York, USA.

Buckland, S. T., D. R. Anderson, K. P. Burnham, J. L. Laake, D. L. Borchers, and L. Thomas. 2004. Advanced distance sampling: estimating abundance of biological populations. Oxford University Press, New York, New York, USA.

Burnham, K. P. 1981. Summarizing remarks: environmental influences. Studies in Avian Biology 6:324–325.

Burnham, K. P., and D. R. Anderson. 1998. Model selection and inference: a practical information-theoretic approach. Springer, New York, New York, USA.

Caughley, G. 1977. Analysis of vertebrate populations. John Wiley and Sons, New York, New York, USA.

Cleary, E. C., and R. A. Dolbeer. 2005. Wildlife hazard management at airports. Second edition. U.S. Department of Transportation, Federal Aviation Administration, Office of Airport Safety and Standards, Washington, D.C., USA.

Cochran, W. G. 1977. Sampling techniques. Third edition. John Wiley and Sons, New York, New York, USA.

Conover, M. R. 2002. Resolving human–wildlife conflicts. CRC Press, Boca Raton, Florida, USA.

DeStephano, S., and R. M. DeGraaf. 2003. Exploring the ecology of suburban wildlife. Frontiers in Ecology and the Environment 1:95–101.

DeVault, T. L., J. L. Belant, B. F. Blackwell, J. A. Martin, J. A. Schmidt, and L. W. Burger Jr. 2012. Airports offer unrealized potential for alternative energy production. Environmental Management 49:517–522.

DeVault, T. L., J. L. Belant, B. F. Blackwell, and T. W. Seamans. 2011. Interspecific variation in wildlife hazards to aircraft: implications for airport wildlife management. Wildlife Society Bulletin 35:394–402.

Diefenbach, D. R., M. R. Marshall, J. A. Mattice, and D. W. Brauning. 2007. Incorporating availability for detection in estimates of bird abundance. Auk 124:96–106.

Dolbeer, R. A. 2006. Height distribution of birds as recorded by collisions with civil aircraft. Journal of Wildlife Management 70:1345–1350.

Dolbeer, R. A., J. L. Belant, and J. L. Sillings. 1993. Shooting gulls reduces strikes with aircraft at John F. Kennedy International Airport. Wildlife Society Bulletin 21:442–450.

Dolbeer, R. A., and S. E. Wright. 2009. Safety management systems: how useful will the FAA National Wildlife Strike Database be? Human–Wildlife Conflicts 3:159–170.

Dolbeer, R. A., S. E. Wright, M. J. Begier, and J. Weller. 2010. Wildlife strikes to civil aircraft in the United States, 1990–2009. Serial Report No. 16. U.S. Department of Transportation, Federal Aviation Administration, Office of Airport Safety and Standards, Washington, D.C., USA.

Dolbeer, R. A., S. E. Wright, and E. C. Cleary. 2000. Ranking the hazard level of wildlife species to aviation. Wildlife Society Bulletin 28:372–378.

Ellingson, A. R., and P. M. Lukacs. 2003. Improving methods for regional landbird monitoring: a reply to Hotto and Young. Wildlife Society Bulletin 31:896–902.

Emlen, J. T. 1977. Estimating breeding season bird densities from transect counts. Auk 94:455–468.

Engeman, R. M., B. U. Constantin, K. S. Gruver, and C. Rossi. 2009. Managing predators to protect endangered species and promote their successful reproduction. Pages 171–187 in A. M. Columbus and L. Kuznetsov, editors. Endangered species: new research. Nova Science, Hauppauge, New York, USA.

Engeman, R. M., R. E. Martin, H. T. Smith, J. Woolard, C. K. Crady, S. A. Shwiff, B. Constantin, M. Stahl, and J. Griner. 2005. Dramatic reduction in predation on marine turtle nests through improved predator monitoring and management. Oryx 39:318–326.

FAA. Federal Aviation Administration. 2004a. Title 14 U.S. Code of Federal Regulations. Part 139: certification of airports. U.S. Department of Transportation, Washington, D.C., USA.

FAA. Federal Aviation Administration. 2004b. Hazardous wildlife attractants on or near airports. Advisory Circular 150/5200-33A. U.S. Department of Transportation, Washington, D.C., USA.

FAA. Federal Aviation Administration. 2007. Hazardous wildlife attractants on or near airports. Advisory Circular 150/5200-33B. U.S. Department of Transportation. Washington, D.C., USA.

Farnsworth, G. L., K. H. Pollock, J. D. Nichols, T. R. Simons, J. E. Hines, and J. R. Sauer. 2002. A removal model for estimating detection probabilities from point-count surveys. Auk 119:414–425.

Graham, R. L., C. T. Hunsaker, R. V. O'Neil, and B. L. Jackson. 1991. Ecological risk assessment at the regional scale. Ecological Applications 1:196–206.

Hoover, S. L., and M. L. Morrison. 2005. Behavior of red-tailed hawks in a wind turbine development. Journal of Wildlife Management 69:150–159.

Hutto, R. L., S. M. Pletschet, and P. Hendricks. 1986. A fixed-radius point count method for nonbreeding and breeding season use. Auk 103:593–602.

Link, W. A., and J. B. Sauer. 1998. Estimating population change from count data: application to the North American Breeding Bird Survey. Ecological Applications 8:258–268.

Lynch, J. F. 1995. Effects of point count duration, time-of-day, and aural stimuli on detectability of migratory and resident bird species in Quintana Roo, Mexico. Technical Report PSW-GTR-149. U.S. Department of Agriculture, Forest Service, Albany, California, USA.

MacKenzie, D. I. 2005. What are the issues with presence-absence data for wildlife managers? Journal of Wildlife Management 69:849–860.

MacKenzie, D. I., J. D. Nichols, G. B. Lachman, S. Droege, A. Royle, and C. A. Langtimm. 2002. Estimating site occupancy rates when detection probabilities are less than one. Ecology 83:2248–2255.

MacKenzie, D. I., and A. Royle. 2005. Designing occupancy studies: general advice and allocating survey effort. Journal of Applied Ecology 42:1105–1114.

Martin, J. A., J. L. Belant, T. L. DeVault, L. W. Burger Jr., B. F. Blackwell, S. K. Riffell, and G. Wang. 2011. Wildlife risk to aviation: a multi-scale issue requires a multi-scale solution. Human–Wildlife Interactions 5:198–203.

Messmer, T. A., S. M. George, and L. Cornicelli. 1997. Legal considerations regarding lethal and nonlethal approaches to managing urban deer. Wildlife Society Bulletin 25:424–429.

Morrison, M. L., W. M. Block, M. D. Strickland, B. A. Collier, and M. J. Peterson. 2008. Wildlife study design. Second edition. Springer, New York, New York, USA.

Nichols, J. D., J. E. Hines, J. R. Sauer, F. W. Fallon, J. E. Fallon, and P. J. Heglund. 2000. A double-observer approach for estimating detection probability and abundance from point counts. Auk 117:393–408.

Reiter, D. K., M. W. Brunson, and R. H. Schmidt. 1999. Public attitudes toward wildlife damage management and policy. Wildlife Society Bulletin 27:746–758.

Reynolds, R. T., J. M. Scott, and R. A. Nussbaum. 1980. A variable circular-plot method for estimating bird numbers. Condor 82:309–313.

Rosenstock, S. S., D. R. Anderson, K. M. Giesen, T. Leukering, and M. F. Carter. 2002. Landbird counting techniques: current practices and an alternative. Auk 119:46–53.

Royle, J. A., and J. D. Nichols. 2003. Estimating abundance from repeated presence-absence data or point counts. Ecology 84:777–790.

Runge, M. C., J. R. Sauer, M. L. Avery, B. F. Blackwell, and M. D. Koneff. 2009. Assessing allowable take of migratory birds: black vultures in Virginia. Journal of Wildlife Management 73:556–565.

Sauer, J. R., J. E. Hines, and J. Fallon. 2008. The North American Breeding Bird Survey: results and analysis, 1966–2007. Version 5.15.2008. U.S. Geological Survey, Patuxent Wildlife Research Center, Laurel, Maryland, USA.

Schafer, L. M., B. F. Blackwell, and M. A. Linnell. 2007. Quantifying risk associated with potential bird–aircraft collisions. Pages 56–63 in C. L. Irwin, D. Nelson, and K. P. McDermott, editors. Proceedings of the 2007 international conference on ecology and transportation. Center for Transportation and the Environment, North Carolina State University, Raleigh, USA.

Scott, J. M., and F. L. Ramsey. 1981. Length of count period as a possible source of bias in estimating bird densities. Studies in Avian Biology 6:409–413.

Seamans, T. W., S. E. Clemons, and A. L. Gosser. 2009. Observations of neck-collared Canada geese near John F. Kennedy International Airport, New York. Human–Wildlife Conflicts 3:242–250.

Stolen, E. D. 2000. Foraging behavior of vultures in central Florida. Florida Field Naturalist 28:173–181.

Thompson, W. L. 2002. Towards reliable bird surveys: accounting for individuals present but not detected. Auk 119:18–25.

15 Conclusions and Future Directions

Jerrold L. Belant
Travis L. DeVault
Bradley F. Blackwell

Although the management of wildlife at airports has seen great progress in recent decades, wildlife collisions with aircraft continue to pose risks to human safety and economic losses to the aviation industry and military (Allan 2002, Dolbeer 2009). Our understanding of physiological and behavioral responses of wildlife to various types of repellents and harassment techniques has grown tremendously. Substantial inroads have been made in developing and optimizing exclusion devices, particularly for mammals. Research and management have increased considerably in recent years, allowing us to better understand aspects of resource use (e.g., cover, food) by wildlife and the spatial scales at which they operate (Martin et al. 2011), as well as to improve current management strategies. We suggest that these two forms of management—repellents and harassment (e.g., Chapters 2–4) and habitat management (e.g., Chapters 8–11)—should be integrated to reduce hazardous wildlife use of airports. Direct control methods (e.g., hazing) typically work only in the short term; reducing habitat suitability for wildlife at airports will likely enhance long-term efficacy of these techniques.

As the integration of several control techniques can result in marked reductions of wildlife use at airports compared to using individual control techniques (see Conover 2002), our improved understanding of ecological theory related to wildlife use of these areas also can enhance our ability to manage associated wildlife risks. Understanding the mechanisms, or causes, of wildlife use of areas at and near airports allows us to

better manage potential hazards. This fundamental mechanistic understanding results in more accurate selection of management options and long-term efficacy of management, which reduces its overall costs. To re-emphasize a simple but effective example, consider a situation described by Bernhardt et al. (2009), who noted comparatively high rates of aircraft collisions with tree swallows (*Tachycineta bicolor*) during autumn at John F. Kennedy International Airport, New York, New York, USA. Rather than increasing harassment actions each autumn to disperse the swallows, airport personnel conducted a study on food resources (Chapter 8) and found that their diet consisted predominantly of northern bayberry fruit (*Myrica pensylvanica*). Determined to be the mechanism or cause of the problem, the bayberry shrubs were subsequently removed. Aircraft strikes with swallows declined markedly in years following bayberry removal, which resulted in reduced hazards to aircraft and allowed airport biologists to focus on other issues.

Although considerable progress has been made in reducing wildlife hazards to aircraft, several important needs for additional information remain. There is need for better understanding of which wildlife species collide most often with aircraft. In the USA, reporting wildlife–civil aircraft strikes to the Federal Aviation Administration (FAA) is voluntary (Cleary and Dolbeer 2005). Heightened public awareness of wildlife collisions with aircraft increased following the crash of US Airways Flight 1549 into the Hudson River (Marra et al. 2009), which in turn increased reporting rates, but only an estimated 39% of all strikes with U.S.-registered aircraft are reported to the FAA (Dolbeer 2009). In addition, only about 26% of reports of wildlife strikes with civil aircraft identify the species involved (Dolbeer and Wright 2009). An improved understanding of the species involved in aircraft collisions could advance our knowledge of those most hazardous to aircraft, as well as strike timing and areas of greatest risk. This knowledge could then help inform airport biologists and contribute to regional- or national-level assessments of risk.

Standardization of survey and monitoring techniques is similarly necessary to ensure consistency in data collection and to allow comparison of hazards at a given airport over seasons or years, as well as to compare relative hazards among airports. In the USA, passenger-certificated airports that experience wildlife hazards are required by the FAA to obtain a Wildlife Hazard Assessment, followed by implementation of a Wildlife Hazard Management Plan (Dolbeer and Wright 2009). Chapter 14 provides a framework that modifies common bird survey approaches to facilitate standardization of data collected within and across airports. One advantage of this approach is the ability to estimate relative species abundance by incorporating imperfect detection of individuals (e.g., MacKenzie 2005). Such standardization and objective-driven data collection can facilitate the development of spatially explicit risk models for airports. Monitoring wildlife use of airports in this manner can improve our ability to discern the best management approaches and to assess the effects of management practices.

An important research emphasis is the development of improved models for estimating risk associated with aircraft collisions, especially for birds. A number of models have been developed in recent years in an effort to quantify risk (Allan 2006; Schafer et al. 2007; Soldatini et al. 2010, 2011). Each of these models in various forms integrates some element of species' relative hazard to aircraft (DeVault et al. 2011), often based in part on body mass (e.g., Allan 2006), as well as abundance and distributions of wildlife species at and near airports. These models are an important step toward assessing wildlife hazards to aircraft, although they pose one apparent disadvantage—they are generally linked to the entire airport and do not adequately consider potential variation in wildlife use of space. Some models

(e.g., Soldatini et al. 2011) consider temporal variation in wildlife hazards, however. Birds typically move in three-dimensional space across time; the importance of considering their altitudinal flight behavior has long been recognized (Major and Dill 1978, DeVault et al. 2005, Avery et al. 2011) and can markedly affect collision rates with aircraft (e.g., Dolbeer 2006). The development of three-dimensional models of birds' probabilistic use of space in relation to aircraft would be a major advancement in risk assessment (Schafer et al. 2007, Belant et al. 2012). For example, habitats surrounding approach and takeoff routes for some airports could be modified on the basis of estimated occurrence of hazardous birds to reduce the probability of collisions.

Advancements in wildlife management at airports have certainly resulted in a reduction of hazardous wildlife at airports (Dolbeer 2011); however, continued and improved efforts are required to minimize suitability of habitats at airports and surrounding areas to wildlife. By continuing to integrate multiple techniques based on the principles of wildlife ecology, and by incorporating technologies that improve our understanding of wildlife and the hazards they pose to aircraft, we can continue to reduce the potential risk of wildlife incidents with aircraft. We cannot ignore new technologies and practices that limit resource availability to wildlife using airports (e.g., DeVault et al. 2012; Chapters 10 and 11). Integration of science with management, through application of new knowledge into airport-specific and national-level guidelines, will further improve the safety of air passengers and reduce economic and biological losses.

Airport managers have long recognized the need and potential advantages of incorporating multiple uses at airports (Infanger 2010), including improved public perception, environmental friendliness (e.g., reducing carbon footprint), and economic incentives. Conserving grassland bird species may be appropriate for some airports (Kelly and Allan 2006), but a lack of scientific data precludes the development of management strategies to conserve grassland birds appropriate for airports (Blackwell et al. 2013). Similarly, increasing global energy demand has resulted in myriad new technologies and applications of alternative energy sources. Although energy production is typically detrimental to wildlife, airports offer one of the few socially acceptable land uses where wildlife use is generally discouraged. Consequently, recent progress has been made in assessing and developing alternative energy sources at airports, especially solar energy (FAA 2010, Infanger 2010, DeVault et al. 2012). Herbaceous biofuels also have potential application at airports, but wildlife use of these plantings and the associated risk to aircraft is less understood than other alternative energy sources (DeVault et al. 2012; Chapter 11).

Integrating management methods that effectively exploit animal sensory capabilities and behaviors, use of resources, movement patterns, and other aspects of animal ecology is vital for reducing wildlife risks to aviation. With an improved understanding of ecological theory and principles as related to wildlife use of airports, airport managers and wildlife biologists can further reduce the number of wildlife–aircraft collisions. It is our hope that this book has provided the basis for such an understanding, and that it will contribute to successful management of wildlife at and near airports worldwide.

LITERATURE CITED

Allan, J. 2002. The costs of bird strikes and bird strike prevention. Pages 147–155 in L. Clark, editor. Proceedings of the National Wildlife Research Center symposium. Human conflicts with wildlife: economic considerations. U.S. Department of Agriculture, National Wildlife Research Center, Fort Collins, Colorado, USA.

Allan, J. 2006. A heuristic risk assessment technique for bird-strike management at airports. Risk Analysis 26:723–729.

Avery, M. L., J. S. Humphrey, T. S. Daughtery, J. W. Fischer, M. P. Milleson, E. A. Tillman, W. E. Bruce, and W. D. Walter. 2011. Vulture flight behavior and implications for aircraft safety. Journal of Wildlife Management 75:1581–1587.

Belant, J. L., J. J. Millspaugh, J. A. Martin, and R. A. Gitzen. 2012. Multi-dimensional space use: the final frontier. Frontiers in Ecology and Environment 10:11–12.

Bernhardt, G. E., Z. J. Patton, L. Kutschbach-Brohl, and R. A. Dolbeer. 2009. Management of bayberry in relation to tree-swallow strikes at John F. Kennedy International Airport, New York. Human–Wildlife Conflicts 3:237–241.

Blackwell, B. F., T. W. Seamans, P. M. Schmidt, T. L. DeVault, J. L. Belant, M. J. Whittingham, J. A. Martin, and E. Fernández-Juricic. 2013. A framework for managing airport grasslands and birds amidst conflicting priorities. Ibis 155:199–203.

Cleary, E. C., and R. A. Dolbeer. 2005. Wildlife hazard management at airports: a manual for airport operators. Second edition. Federal Aviation Administration, Office of Airport Safety and Standards, Washington, D.C., USA.

Conover, M. R. 2002. Resolving human–wildlife conflicts: the science of wildlife damage management. CRC Press, Boca Raton, Florida, USA.

DeVault, T. L., J. L. Belant, B. F. Blackwell, J. A. Martin, J. A. Schmidt, and L. Wes Burger Jr. 2012. Airports offer unrealized potential for alternative energy production. Environmental Management 49:517–522.

DeVault, T. L., J. L. Belant, B. F. Blackwell, and T. W. Seamans. 2011. Interspecific variation in wildlife hazards to aircraft: implications for airport wildlife management. Wildlife Society Bulletin 35:394–402.

DeVault, T. L., B. D. Reinhart, I. L. Brisbin Jr., and O. E. Rhodes Jr. 2005. Flight behavior of black and turkey vultures: implications for reducing bird–aircraft collisions. Journal of Wildlife Management 69:601–608.

Dolbeer, R. A. 2006. Height distribution of birds recorded by collisions with civil aircraft. Journal of Wildlife Management 70:1345–1350.

Dolbeer, R. A. 2009. Trends in wildlife strike reporting, part 1: voluntary system, 1990–2008. Report DOT/FAA/AR-09/65. U.S. Department of Transportation, Federal Aviation Administration, Washington, D.C., USA.

Dolbeer, R. A. 2011. Increasing trend of damaging bird strikes with aircraft outside the airport boundary: implications for mitigation measures. Human–Wildlife Interactions 5:235–248.

Dolbeer, R. A., and S. E. Wright. 2009. Safety management systems: how useful will the FAA National Wildlife Strike Database be? Human–Wildlife Conflicts 3:167–178.

FAA. Federal Aviation Administration. 2010. Technical guidance for evaluating selected solar technologies on airports. Report FAA-ARP-TR-10-1. Washington, D.C., USA.

Infanger, J. F. 2010. The pros, cons of solar, wind. Airport Business 24:18–19.

Kelly, T., and J. Allan. 2006. Ecological effects of aviation. Environmental Pollution 10:5–24.

MacKenzie, D. 2005. What are the issues with presence-absence data for wildlife managers? Journal of Wildlife Management 69:849–860.

Major, P. F., and L. M. Dill. 1978. The three-dimensional structure of airborne bird flocks. Behavioral Ecology and Sociobiology 4:111–122.

Marra, P. P., C. J. Dove, R. A. Dolbeer, N. F. Dahlan, M. Heacker, J. F. Whatton, N. E. Diggs, C. France, and G. A. Henkes. 2009. Migratory Canada geese cause crash of US Airways Flight 1549. Frontiers in Ecology and the Environment 7:297–301.

Martin, J. A., J. L. Belant, T. L. DeVault, B. F. Blackwell, L. W. Burger Jr., S. K. Riffell, and G. Wang. 2011. Wildlife risk to aviation: a multi-scale issue requires a multi-scale solution. Human–Wildlife Interactions 5:198–203.

Schafer, L. M., B. F. Blackwell, and M. A. Linnell. 2007. Quantifying risk associated with potential bird–aircraft collisions. Pages 56–63 *in* C. L. Irwin, D. Nelson, and K. P. McDermott, editors. Proceedings of the 2007 international conference on ecology and transportation. Center for Transportation and the Environment, North Carolina State University, Raleigh, USA.

Soldatini, C., Y. V. Albores-Barajas, T. Lovato, A. Andreon, P. Torricelli, A. Montemaggiori, C. Corsa, and V. Georgalas. 2011. Wildlife strike risk assessment in several Italian airports: lessons from BRI and a new methodology implementation. PLoS ONE 6:e28920. doi:10.1371/journal.pone .0028920.

Soldatini, C., V. Georgalas, P. Torricelli, and Y. V. Albores-Barajas. 2010. An ecological approach to birdstrike risk analysis. European Journal of Wildlife Research 56:623–632.

Appendix

Regulations for Wildlife Management at Airports

RICHARD A. DOLBEER

In 1990, the 190 member nations of the International Civil Aviation Organization (ICAO) adopted, in Annex 14 to the Convention on Civil International Aviation, three recommended management practices regarding bird hazards to aviation. The recommended practices required that aviation authorities within each nation (1) assess the extent of the hazard posed by birds at and in the vicinity of airports certificated for passenger traffic, (2) take necessary action to decrease the number of birds, and (3) eliminate or prevent the establishment of any site in the vicinity of the airport that could attract birds and thereby present a danger to aviation. Because of the increasing threat posed by birds to aviation worldwide, member states voted to make these recommendations mandatory ICAO standards, effective November 2003. In 2009, ICAO expanded these standards to include terrestrial wildlife such as large mammals and reptiles that pose a risk at airports (ICAO 2009).

To comply with ICAO standards, the Federal Aviation Administration (2004) requires airports in the USA that are certificated for passenger traffic to conduct Wildlife Hazard Assessments (WHAs). In 2012, there were 550 such airports in the USA. General aviation airports receiving federal funding also may be required to conduct WHAs. Based on the findings of the WHA, most airports are required to develop and implement a Wildlife Hazard Management Plan (WHMP). These WHMPs, as dictated by requirements outlined in Federal Aviation Administration (2004), must address (1) removal of habitat and food attractive to wildlife; (2) use of techniques to exclude, disperse, or remove wildlife that pose a risk to aircraft; (3) training of airport personnel in wildlife management techniques; and (4) establishment of an Airport Wildlife Hazard Working Group. In addition, Federal Aviation Administration (2007) provides guidance on land uses that attract hazardous wildlife and are thus incompatible with aviation safety within 13 km (8 miles) of aircraft operating surfaces at airports (e.g., garbage landfills).

In implementing WHMPs in the USA, airports must deal with numerous regulatory constraints related to environmental issues at the federal, state, and local levels. First, most birds are federally protected under the Migratory Bird Treaty Act (MBTA), which is administered by the U.S. Department of Interior, Fish and Wildlife Service (USFWS). The MBTA (16 U.S.C. § 703–712) is an international treaty signed by the USA, Mexico, Canada, Russia, and Japan (USFWS 2013). State laws protecting birds can be even more (but not less) restrictive. State laws regulate most other wildlife, including mammals and reptiles. Before any management action can be taken to kill, trap and translocate, or disrupt reproduction of any species covered by these laws, federal and state permits must be obtained. These permits dictate allowable management methods and the numbers of animals or eggs that can be removed by species. The management of wildlife species classified as endangered has additional constraints under the Endangered Species Act (16 U.S.C. § 1531–1544), also administered by the Department of Interior. Most states have their own endangered species legislation, as well.

A constraint in implementing WHMPs at many airports relates to the management of wetlands that are attractive to birds. Under Section 404 of the Clean Water Act of 1972 (33 U.S.C. § 1251 and related legislation), most wetlands cannot be removed without obtaining a permit from the U.S. Army Corps of Engineers. These permits typically require that any removed or negatively altered wetland must be mitigated by the establishment of wetlands in other locations within the same watershed.

The U.S. Environmental Protection Agency (2012) oversees, through the Federal Insecticide, Fungicide and Rodenticide Act (7 U.S.C. § 136.40), the use of pesticides such as chemical toxicants and repellents that may be used to manage wildlife at airports. All pesticides must be registered with the Environmental Protection Agency as either general or restricted use before they can be applied under specified label directions. Restricted-use pesticides can only be applied by state-licensed applicators. Finally, state and local regulations may also constrain the use of firearms, traps, and bird-frightening devices that emit loud noises (e.g., propane cannons).

At international and national levels, the ICAO and civil aviation authorities, respectively, mandate that airports assess and manage the risk caused by birds and other wildlife in the airport environment. This task is made uniquely challenging, however, as described above, by the numerous constraints imposed by complex environmental regulations overseen by various federal, state, and local agencies. These environmental regulations, although unquestionably beneficial for society as a whole, are often at cross-purposes with aviation safety in the airport environment. The FAA and U.S. Department of Agriculture have published a 348-page manual, *Wildlife Hazard Management at Airports* by Cleary and Dolbeer (2005), which provides detailed guidance and background material for personnel conducting WHAs and implementing WHMPs in relation to environmental regulations.

Literature Cited

Cleary, E. C., and R. A. Dolbeer. 2005. Wildlife hazard management at airports: a manual for airport personnel. Second edition. U.S. Department of Transportation, Federal Aviation Administration, Office of Airport Safety and Standards. Washington, D.C., USA.

Federal Aviation Administration. 2004. Title 14 U.S. Code of Federal Regulations. Part 139: certification of airports. U.S. Department of Transportation, Washington, D.C., USA.

Federal Aviation Administration. 2007. Hazardous wildlife attractants on or near airports. Advisory Circular 150/5200-33B. U.S. Department of Transportation, Washington, D.C., USA.

ICAO. International Civil Aviation Organization. 2009. Convention on international civil aviation (standards and recommended practices). Annex 14: aerodromes. Volume 1. Aerodrome design and operations. Sixth edition. Montreal, Quebec, Canada.

U.S. Environmental Protection Agency. 2012. Summary of the Federal Insecticide, Fungicide, and Rodenticide Act. Washington, D.C., USA. http://www.epa.gov/regulations/laws/fifra.html.

USFWS. U.S. Fish and Wildlife Service. 2013. Digest of federal resource laws. Washington, D.C., USA. http://www.fws.gov/laws/lawsdigest/Resourcelaws.html.

Index

Page numbers in *italics* indicate figures and tables.